ELECTRO-OPTICAL EQUIPMENT IN MECHANICAL HANDLING AND SORTING

ANDERSONIAN LIBRARY

WITHDRAWN FROM LIBRARY STOCK

University of STRATHCLYDE

WITHDRAWN
FROM
LIBRARY
STOCK

ELECTRO-OPTICAL EQUIPMENT IN MECHANICAL HANDLING AND SORTING

Conference sponsored by the
Postal Engineering Section of the Manipulative and
Mechanical Handling Machinery Group and co-sponsored
by the Institution of Electronic and Radio Engineers

London, 7 May 1974

Published by
Mechanical Engineering Publications Limited for
The Institution of Mechanical Engineers
LONDON AND NEW YORK

CP12 - 1974

UNIVERSITY OF
STRATHCLYDE LIBRARIES

First published 1975

This Publication is copyright under the Berne Convention and the International Copyright Convention. Apart from any fair dealing for the purpose of private study, research, criticism or review as permitted under *The Copyright Act 1956*, no part may be reproduced, stored in a retrieval system or transmitted in any form or by any means, electronic, electrical, chemical, mechanical, photocopying, recording or otherwise, without the prior permission of the copyright owners. Inquiries should be addressed to J. C. Mundy, Publications, Projects, and Liaison Officer, The Institution of Mechanical Engineers, 1 Birdcage Walk, Westminster, London, SW1H 9JJ.

© The Institution of Mechanical Engineers 1974

This volume is complete in itself.
There is no supplementary discussion volume.

ISBN 0 85298 307 7

UNIVERSITY OF
STRATHCLYDE
DAMAGED
181289

Set and printed photolitho by Gwynne Printers Ltd, Hurstpierpoint, Sussex.
Bound by F J Blisset & Co., Palmerston Works, Roslin Road, Acton, London W3

Made and printed in Great Britain

D
621.86
ELE

CONTENTS

C122/74

INTEGRATED OPTO-ELECTRONICS IN INDUSTRIAL ENVIRONMENTS

PETER BARTLAM B Sc
Chief Engineer (Instruments), Integrated Photomatrix Limited,
The Grove Trading Estate, Dorchester, Dorset
The Ms. of this paper was received at the Institution on 19th February 1974 and accepted for publication on 28th February 1974.

SYNOPSIS   The paper describes the principle of operation of linear and two-dimensional self-scanned silicon photodiode arrays and their application to industrial measurement and inspection problems. Specific systems for flaw detection and dimension gauging are used for illustration. Other applications and potential applications are discussed and a brief indication of system design parameters is outlined.

INTRODUCTION

The rapid progress of silicon micro-electronic technology in recent years has resulted in the development of opto-electronic devices in which the photo-sensor is manufactured on the same integrated circuit chip as the associated processing electronics. These technology advances have in turn resulted in the application of silicon detectors to a wide range of problems which had hitherto proved impractical using discrete photodiodes and associated circuitry. The semiconductor process used is the metal-oxide-semiconductor (MOS) process which was originally developed for large scale integrated (LSI) digital circuits, e.g. semiconductor memories. The MOS process is ideally suited in particular to producing self-scanned arrays of photodiodes in which the outputs of the photodiodes are accessed onto a single output pin by an MOS shift register. The specific advantages of MOS technology are high input impedance switch characteristics, low power consumption and high packing density.

These arrays, originally developed for optical character recognition (OCR), have found many diverse applications in the field of measurement and inspection and this paper will outline some of the more significant applications in these areas.

The Light Integration Mode

When integrated with MOS circuitry the silicon photodiode is operated in the light integration mode, whereby the self-capacitance of the diode is charged, by an MOS transistor to a negative potential thus reverse biasing the diode. When this transistor is turned off the photodiode capacitance sees an extremely high impedance and the decay of charge is determined by carriers produced by the incident light falling on the photodiode. Fig.1 shows a basic element consisting of a photodiode and an MOS transistor. Without any light falling on the diode, there is a decay of charge due to the leakage current of the diode and this limits the minimum light level at which the technique may be used.

The typical responsivity of the photodiode operated in the light integration mode is $25V \, sec^{-1}\mu W^{-1}cm^2$, i.e. this bias decay would be 25V in an integration period of 1 second for an incident light level of $1\mu Wcm^2$. Since decay is essentially linear then it follows that an integration time of 10mS would produce a signal of 250mV. The 1 second integration period is largely theoretical because of the leakage current effect. Responsivity is specified for the light integration mode as signal output per unit of exposure, where exposure is defined as the light intensity multiplied by the integration time.

The Self-Scanned Linear Array

The simplest form of photodiode array is a linear array of photodiodes in which the multiplexing of the MOS transistors shown in Fig.1 is carried out by a shift register integrated alongside the line of photodiodes. The shift register accesses video information from each photodiode sequentially as a logic '1' is propagated through the register. There are two modes of operation used for accessing video information from self-scanning arrays. In both cases, each photodiode is charged to a negative reference voltage through an MOS switch as indicated in Fig.1. From the time that this switch is turned off, the charge held on the diode, owing to its self-capacitance, will decay slowly as a result of (a) its self leakage current and (b) a photocurrent generated by incident-illumination. The leakage current is insignificant except at very low levels, and the charge loss is directly proportional to the incident illumination.

In the first system the diode is sampled immediately prior to being charged to the reference voltage. This is achieved with an MOS source follower and sample switch and produces an output voltage proportional to the diode voltage at the end of the integration period. The mode of processing the video signal is thus termed the voltage sampling mode. Fig. 2 illustrates one element of a voltage sampled array where shift register output 'n' operates the sample switch and the

1

following output 'n+1' operates the switch to recharge the diode to the reference potential.

In the second mode used the diode is again recharged to the reference voltage, and the amount of charge required to reinstate the original voltage appears as a current pulse on the supply line. Charge amplification, together with complex integration and sample-and-hold functions are necessary to produce the more familiar boxcar waveform output similar to that produced in the voltage-sampling mode. This mode is thus termed the recharge-sampling mode and one element of such an array is as shown in Fig.1. The extra complexity of signal processing employed in the recharge mode results in a superior performance specifically in high multiplexing rates (up to 5 MHz), uniformity of response and signal/noise ratio. These features are of particular importance in OCR applications where a number of grey levels must be resolved, but in the majority of industrial applications involving the detection of flaws or edges the use of the voltage-sampling mode with its inherent simplicity of signal processing is often advantageous. Typical self-scanned linear arrays available are 16, 32, 50, 64, 100, 128 and 256 element arrays with a centre-centre spacing of .004 ins. while the state-of-the art development is a linear array of 1024 diodes on a centre-centre spacing of .001 ins. Fig. 3 shows two typical self-scanned linear arrays.

The technique as developed on linear arrays has now been extended to 2-dimensional matrix arrays where accessing of the photodiodes is carried out by two MOS shift registers. The recharge method of processing is used. The state-of-the-art on MOS scanned 2-dimensional arrays is a 64 x 64 (i.e. 4096 diodes) array with a centre-centre spacing of .003 ins.

The linear self-scanned array has now been in existence for several years and there are many users with considerable experience of their application to industrial problems. These applications tend to be diverse and specialised and it is often the case that the potential user either does not possess the necessary electronics expertise or cannot afford the development time involved to engineer such systems to meet their specific requirements. In addition, the interfacing of an array often requires a knowledge of optical systems to achieve optimum performance. For this reason, camera systems have been developed which provide the basic electronics, hardware and optics to form the basis of an engineered system for any given application.

The same approach has been taken also with the development of 2-dimensional arrays though here applications must be considered as potential due to the fact that the device is a recent development.

Flaw Detection

Flaw detection is a problem in many manufacturing industries, particularly where a high volume of piece-parts is involved or where material is produced in a continuous strip at a high rate. In both cases the use of self-scanned linear arrays for automatic inspection offers obvious advantages. Fig. 4 illustrates the basic system required for on-line detection of flaws, e.g. pinholes, in such materials as emulsion-coated tape or film. In this case, the material is illuminated from below so that pinhole flaws will have a high transmission of light giving a high contrast against the emulsion coating. On many other materials such as sheet metal, it is necessary to use a reflection system in order to detect surface blemishes.

Consider an example of film which is coated with an emulsion as a 50 cm wide web prior to further expensive processing. It is necessary to identify sections of the web which contain flaws as small as 1 mm, and this may be achieved using an array camera system as illustrated in Fig.5. The array length required to resolve small pinholes is not as great as would first seem apparent. This is due to the high transmission of light through even a tiny flaw and also to the fact that it is not important to accurately locate the position of such flaws but merely to identify the zones of the web in which flaws occur. In order to detect flaws down to 1 mm dia. across the 50 cm wide web, a camera employing an array of 100 photodiodes is adequate. The lens used is a 16 mm cine lens f/1.9 of 25 mm focal length and to give the correct magnification and resolution, the camera is located 120 cm above the web. To discriminate flaws the video output should be connected to a voltage comparator. Clock pulses and the start-of-scan pulse can be utilised to identify the location of flaws in zones across the web by means of simple logic. Footage information can be produced from a separate sensor which may be a pick-off from the mechanical drive. The output of the logic could conveniently drive a printer to give a print-out of the location of flaws in terms of zones and footage thus permitting sections with flaws to be removed prior to any further processing. Another useful technique in flaw detection of separate items is to simultaneously scan two items, nominally identical, with separate cameras and to compare electronically the video outputs. If a flaw appears on one of the items then a difference signal will result. Assuming that the position of flaws is random then the probability of getting two items with identical flaws is remote. If the manufacturing process is such that there is a likelihood of repetitive flaws occurring, then a master item having no flaws must be used for the comparison. In each case, of course, the alignment of the two cameras must be precise and this technique is currently being used for the detection of flaws in banknotes. The same techniques could be successfully employed for the automatic inspection of items such as plastic or rubber mouldings where the quantities involved are very large and the use of visual inspection techniques necessitates the employment of many trained operators.

Dimension Gauging

The very nature of the self-scanned array, and the method by which video information is digitally accessed, makes the device ideally suited to edge detection where the position of an optical edge can be accurately located along the array. Thus, any part which can be optically imaged onto an array can be gauged by this technique, either as an on-line monitoring function, as in the case of steel, glass, etc., or as a piece-part inspection facility where the parts are fed by a mechanical handling system through the gauging head and then sorted according to measured size.

Inspection systems utilising photodiode arrays generally involve customer engineering and this makes such specialised systems relatively expensive, the final cost depending upon the length of array or arrays employed and the total complexity of the system. The use of such a system must be economically viable, and the application of self-scanned arrays in the field of dimension monitoring becomes justifiable in cases where a high degree of inspection reliability is involved, e.g. mass produced items such as certain automotive parts, or products where the cost of scrap or material processing is relatively high.

Complex gauging systems have been developed where a small process computer has been employed. One example of this type of system is a hot steel rod gauging system employing several separate optical heads at different locations in the mill each feeding measurement data to the process computer for analysis and recording. Fig. 6 shows the block diagram of the basic measurement head and Fig. 7 shows a photograph of one such measurement head where the two arrays can clearly be seen. The essence of the technique is to digitally count the total number of diodes obscured on the two arrays. The principle of width gauging is illustrated using one array in Fig. 8 where a rod can clearly be seen imaged onto an array and the resultant video signal is shown on the oscilloscope.

Another similar example is the use of two separate line-scan cameras for monitoring railway track gauge. In this case the edge of the track is illuminated by a high intensity light source and the worn surface of the rail produces an optical contrast against the remainder of the rail suitable for imaging onto a photodiode array. Again, the total number of diodes obscured is digitally counted, and the data is fed to the computer to analyse and record variations in track gauge.

Yet another dimension gauging system which utilises a small process computer is a sawmill system for measuring and recording length and diameter of logs. The measured data is used to sort logs by size and generate an accurate inventory report at the end of each mill shift. Particular advantages of self-scanned arrays in the above type of gauging system are: non-contacting measurement, high accuracy and repeatability, digital signal processing, on-line capability and fast response time. An additional advantage in the case of width measurement systems using two arrays is that lateral movement does not produce any measurement errors, as the total number of diodes obscured does not vary with lateral motion until the optical edge reaches the end of an array. In fact where a computer is employed the lateral movement of the object may be used to advantage to increase the accuracy or resolution of movement. This is achieved by taking a large number of readings which, due to the fluctuation of the object, will have a statistical spread. The spread is analysed by the computer to give an increase in measurement accuracy proportional to the total number of readings analysed. The technique has been successfully used to improve accuracy by a factor of 100 over the basic static optical accuracy. The basic gauging technique, as illustrated by the above examples, has also been applied to on-line monitoring of hot glass, colour printing registration using reference marks and hole area measurement. There are clearly many other potential applications in on-line measurement and gauging and automatic sorting of piece-parts.

Optical Character Recognition

Self-scanned linear arrays have been extensively used as the sensor for modern OCR systems where documents, such as cheques, are read at a high rate for the automatic input of data into a computer. Such arrays are currently being evaluated for use in systems for postal code reading and pools coupons reading using established OCR techniques. In material handling systems the self-scanned array may be used as the sensor for reading characters or codes on packages.

Applications of the Two-dimensional Array

There is no doubt that the advent of the two dimensional self-scanned array will reveal some unforeseen novel applications, but at the moment it is possible to foresee applications in many of the areas in which the linear arrays are now used. For flaw detection similar techniques to those outlined using linear arrays may be used for automatically checking objects while they are static. Limited resolution arrays with a typical matrix format of 40 x 15 are currently being evaluated for OCR type applications such as in point-of-sale terminals. The matrix size of current 2-D arrays is limited in comparison with vidicon cameras but there are areas where the solid state and precision characteristics of the array may render it advantageous as an imaging device, i.e. a solid state TV camera with a low picture resolution; such an area is in the field of automatic assembly using established pattern recognition techniques where a 2-D array could be used as the sensor to recognise a piece-part and then follow a routine assembly operation defined by a predetermined programme. Ultimately as the technology advances it will be possible to produce a solid state TV camera using such 2-D arrays with standard video resolution.

System Design Considerations

The most significant design consideration for most of the above applications is the optical consideration of element resolution and overall range. This determines the length of array required, e.g. if an element resolution of .010 in. is required over a total measurement range of 1 inch then a 100 element array

is required. In the case of flaw detection the
resolution considerations are dependent upon
the form of the flaw being detected as
described in the relevant section of this paper.
The basic resolution required combined with the
diode element size determines the optical
magnification and hence the type of lens. The
signal output as a function of light level is
directly proportional to the integration time,
i.e. the longer the integration period the
higher the signal level. For most industrial
applications it is possible to use the longest
possible integration period for the arrays
(2 mS) and thus have maximum responsivity.
This must be optimised according to the
required data rate but normally one reading in
2 mS is adequate.

In considering the feasibility of any given
application it is usually necessary to study
the optical details practically in order to
determine the optimum form of illumination to
obtain good optical contrast. The array
limitation in terms of detectability of a given
signal is the inherent signal/noise ratio.
Under normal conditions the limiting noise is
fixed pattern noise which is the variation of
signal output from diode to diode across the
array for a fixed illumination level.

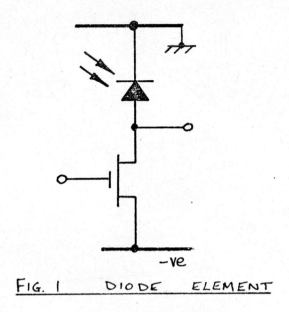

FIG. 1   DIODE ELEMENT

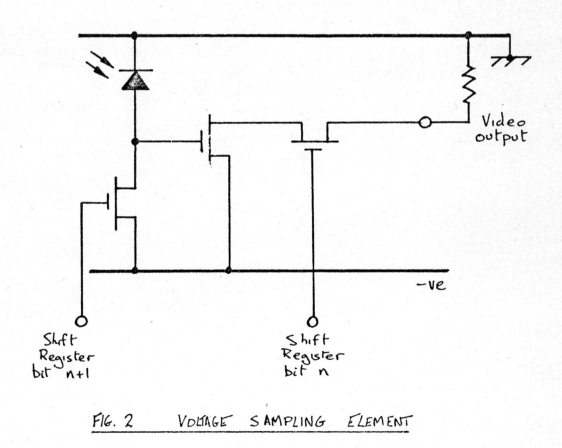

Video output

Shift Register bit n+1

Shift Register bit n

−ve

FIG. 2   VOLTAGE SAMPLING ELEMENT

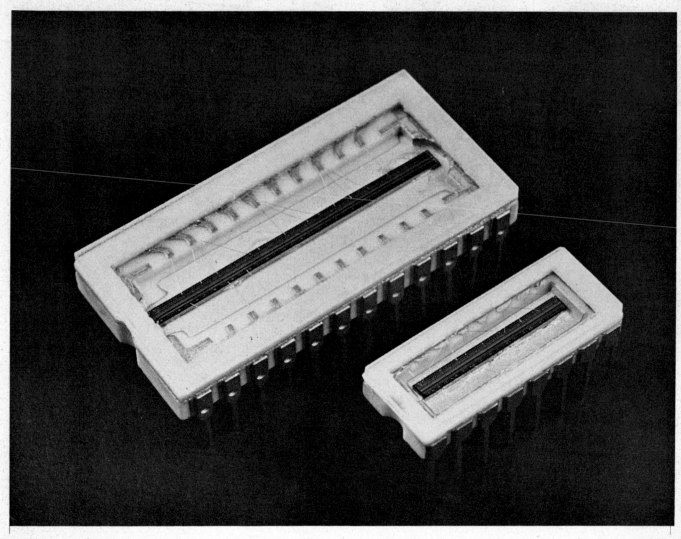

Fig. 3. Typical Linear Arrays

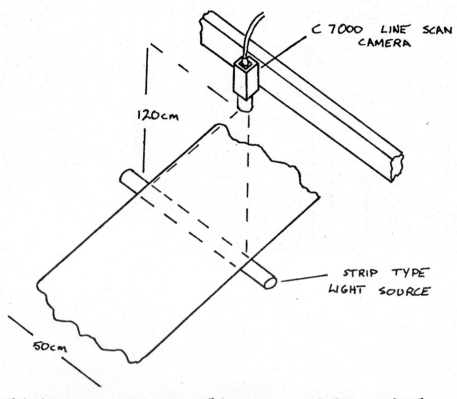

C 7000 LINE SCAN CAMERA

120cm

STRIP TYPE LIGHT SOURCE

50cm

FIG. 4 FLAW DETECTION OF EMULSION COATED TAPE.

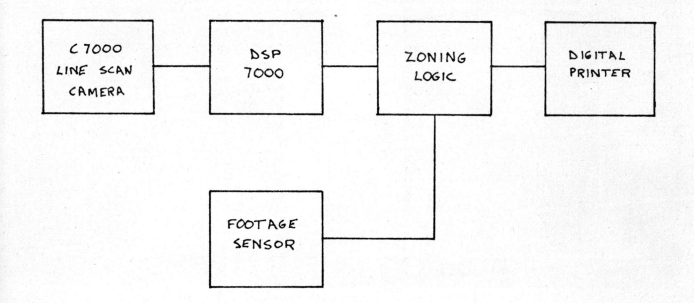

FIG. 5 BLOCK DIAGRAM OF TYPICAL FLAW DETECTION SYSTEM

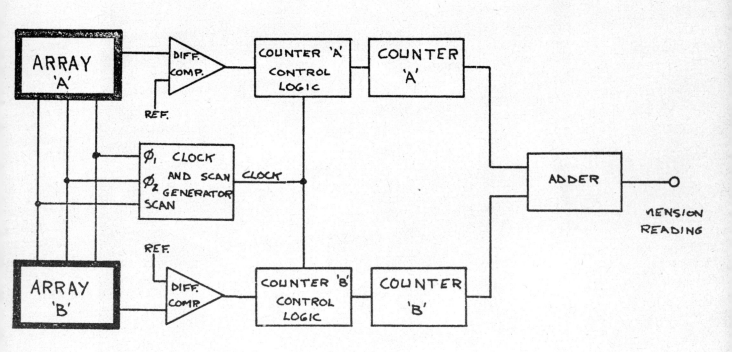

FIG. 6 BLOCK DIAGRAM OF HOT ROD GAUGE MEASUREMENT HEAD.

Fig. 7. Hot Rod Gauge Measurement Head

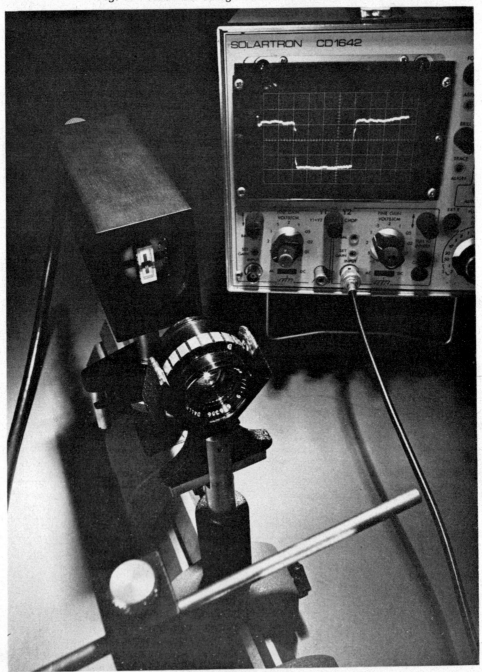

Fig. 8.
Width Gauging
Principle

C123/74

ELECTRO-OPTICAL DEVICES USED IN POSTAL MECHANISATION

T. W. GREATHEAD, C Eng, MIEE
J. L. BUSBY, B Sc
Executive Engineers, Postal Headquarters,
Mechanisation & Buildings Department
The Ms. of this paper was received at the Institution on 18th January 1974 and accepted for publication on 28th February 1974. 44

SYNOPSIS   A review is given of applications of photo-electric equipment in mail handling installations within the British Post Office. Two important new developments are then described; the first enables the unambiguous detection of irregularly shaped parcels, the second employs closed circuit television as a means of presenting letter mail addresses to coding machine operators. Finally, the applications of recently developed devices to postal mechanisation equipment are considered.

1.   INTRODUCTION

For the control of mail during machine processing devices are needed to indicate the position of items separately and in bulk. There are several ways of achieving this, and the Post Office relies mainly on simply techniques such as photo-sensing. Some of the applications demand new and novel electro-optical methods, but most of the equipment is similar to that used by other industrial organisations. This paper illustrates how Postal Engineers have used conventional techniques to solve a large variety of problems and describes some of the new devices that are being developed.

2.   TECHNIQUES

A simple sensing device can be used to detect the presence of letters because they are usually regular and have uniform leading and trailing edges. The detection system can be compact and close to the letter path since machineable letters are not more than $\frac{1}{4}$ inch thick.

Parcels are neither regular nor uniform in size, shape or weight so that where accuracy is important a continuous total area scanning device is necessary to detect the presence of a parcel and provide adequate information about its length and position. The variable width of parcels prevents a close scan.

The basic photo-electric device consists of a light source, which projects a well-defined light beam, a photo-electric sensor mounted in a suitable housing (usually incorporating a focusing lens) and a threshold detection circuit. Interruption of the beam causes the detection circuit output to indicate the presence of a parcel or a letter. The output of the detection circuit is caused to drive a relay for simple electro-mechanical controls or it may drive solid state switching logic when more complex control functions are required. For more sensitive work such as the detection of stamps and phosphor marks on envelopes, photo-multiplier tubes are used. These require sensitive amplifiers, since the output current change is in the order of only 2uA.

Photo-electric devices must operate reliably in the hostile environment of a sorting office. Maintenance problems arise because of dust,

vibration, accidental mechanical damage and interference from extraneous light sources. These effects can be minimised by careful design and by taking precautions during the installation and running of the equipment; for example, problems with dust can be reduced by placing optical surfaces vertically and vibration effects can be lessened by mounting units on rigid surfaces (preferably not part of moving equipment). Also, by underrunning lamps the risk of misalignment due to filament sag is reduced (10% reduction in voltage results in about 30% reduction in light output and can increase lamp life by up to 400%). Masking is used to exclude extraneous light.

For special purposes, more sophisticated photo-electric systems are employed, using a modulated light source, a high gain frequency selective amplifier and a sensing unit in a single positive feedback loop.

3.   TYPICAL APPLICATIONS IN PARCEL & BULK MAIL HANDLING

Whilst, for overall control in large mechanised sorting offices, closed circuit television (CCTV) is being used increasingly to supervise the general flow of parcels, photo-electric devices are used as jam detectors, load level monitors and for machine control functions. Jam detectors are installed at points where there is a known risk of blockage in the mail flow, and load level monitors are used with storage devices to either stop the feed to a full store or to divert it to some other store. Examples of some machine control functions are given below.

3.1 Parcel Conveyor Control

A belt conveyor used for storage (and supplied with parcels from a chute) has two photo-electric units, one of which is operated when the space under the chute is full. When this condition is detected the conveyor moves forward until the space under the chute is empty. The other detector is at the far end of the conveyor positioned level with the bottom layer of parcels. This operates when the conveyor is full and about to discharge. If the control system is set to "Store" this detector overrides the inching control and stops the conveyor, which then becomes a stationary store.

3.2 Letter Tray Conveyors

These are used to convey letters in bulk. When a tray is loaded on to the conveyor the destination code (0-99) is set by arranging small movable reflectors on the front and rear edges of the tray (see Figure 1). At appropriate points in the conveyor system the code is read by means of a photo-sensing head. There are three types of sensing head in use; the code reader which looks for a particular code, the code checker which looks for an incorrectly coded tray and the tray detector which controls the flow of trays. A tray detector may be associated with a code checker so that if a tray with an incorrect code passes a code checker the conveyor on which it resides is stopped.

3.3 Parcel Sorting Machines

The tilted belt parcel sorting machine (PSM) relies upon photo-electric equipment in a number of ways. For example, when the PSM operator reads the address on a parcel he presses the appropriate button to enter the sorting code into an electronic memory and places the parcel on the tilted belt. The parcel travels towards a photo-electric synchronizing head situated 3 to 4 ft from the sorting position and breaks the light beam which causes the sorting code to be transferred to the main memory unit where it travels in synchronism with the parcel on the belt. When the parcel reaches the appropriate sorting outlet a door opens and the parcel slides off the belt into a small capacity store. When a store for a particular destination is full the photo-electric load level monitor operates and the door remains closed. (Any parcels en route to this door are then taken to the overflow outlet where they are stored until they can be resorted). When at least three load level monitors (and/or the overflow monitor) are operated the machine automatically stops.

3.4 Belt Tracking Correction

Foxwell corrector beams (consisting of three control beams situated at the edge of some special purpose conveyor belts) are used to compensate for a sideways shift in the position of the belt caused by parcels being moved across the belt by a diverter. When this happens pinch rollers beneath the belt are used to restore it automatically to its correct position.

4. TYPICAL APPLICATIONS IN LETTER MACHINERY

With the exception of the coding desk, the throughput of which depends on the operator's keying rate, the automatic letter processing machines work at synchronous speeds. Photo-sensing units determine the positions of letters within the transport mechanism and are used to control the appropriate machine functions, eg diverters, cancelling dies, counters, jam alarms, etc. Detailed descriptions of the various machines have been published in other papers; only a brief outline of the relevant features is included here.

4.1 Automatic Facing Machine (ALF)

Letters are exposed briefly to ultra-violet light, which causes the phosphor bars on the stamps to emit visible light in the blue band of the spectrum for up to half a second after leaving the irradiation unit. Photo-multiplier tubes are used to detect this re-radiated light and enable the positions of the stamps (and consequently the orientation of the letters) and the number of phosphor bars per letter to be recorded in a series of shift registers through which the information is stepped in synchronism with the passage of the letters through the machine. This enables the machine to cancel and date-stamp each item and stack it according to its address orientation and tariff rating. Photo-multiplier tubes are also provided to record the change in reflection of a beam of visible light from non-phosphor treated items such as official paid and business reply items.

4.2 Coding Desk

This machine is controlled by an operator using a keyboard to copy-type the address code from the letter. When the keyboard information has been processed and the machine is ready to print the corresponding phosphor code pattern on the envelope the letters waiting at each of the various stages are moved on simultaneously but photo-beams are used to control their respective stopping positions. This allows letters of various lengths and widths to be stopped with their appropriate edges in the correct positions at each stage of the machine.

4.3 Automatic Sorting Machine (MSASM)

Letters, with phosphor code marks applied by the coding desks, are irradiated by ultra-violet light of longer wavelength than that used in the facing machine; this prevents any unwanted signals being generated by stamps which may pass the reading head. Photo-multiplier tubes having a peak response at approximately 430 nm, corresponding to the wavelength of the re-radiated light, are used to detect the phosphorescent code marks. The presence or absence of code marks at $\frac{1}{4}$ inch intervals is converted by a sophisticated strobing circuit into electrical signals which are stored until the complete code pattern has been accumulated. These signals are then converted into routing information which is used to control the passage of the letter through the machine and into the appropriate sorting box. Signal verification and parity checking is used, and any letters with codes not conforming to a standard format are routed to a special box.

4.4 Special Purpose Photo-beam Units

The information stored in shift registers in the facing and sorting machines is transferred from stage to stage by a series of pulses obtained from the interruption of a light beam by a perforated disc driven by the letter transport mechanism. The same technique is used in the coding desk replenisher equipment, and the stacker-canceller machine separates long and short letters by counting such pulses whilst letters travel past a photo-beam, after which they are diverted into stacks according to their relative lengths.

A pair of photo-beam units are used to monitor the apparent pivot corner angle of letters as they pass through a 90° turn unit in the sorting machine. In this way overlapping items can be detected, since the apparent pivot angle increases from 90° to 180°, according to the amount of overlap. The beams are arranged so that a single item cannot obscure both simultaneously; this can only happen if both items overlap to some extent.

5. CURRENT DEVELOPMENT IN PARCEL AND BULK MAIL HANDLING

5.1 The Need for a Parcel Detector

There is a demand in Parcel Mechanisation for a device that measures the length of parcels as they pass by on a conveyor. The primary use for a device of this nature is either on the tilted belt parcel sorting machine (PSM) or on its associated dual feed equipment, where parcels travel separately. If the length of each parcel can be determined accurately, the space between parcels could be reduced and the overall throughput could be increased by about 30%. However, a suitable device would have to deal with a large variety of parcels including those that are irregularly shaped with hollows or bulges such as to register on a single beam detector as more than one parcel. The essential performance requirements for such a device are consequently stringent (see Appendix).

To meet this demand a recent development (which is the subject of a patent application) has produced a versatile device capable of measuring accurately the length (or any other dimensional parameter) of even the most irregularly shaped parcels. This detector is a continuous scanning device capable of detecting the smallest object within its resolution capability anywhere in its field of scan. Moreover, it costs no more than the cheapest system using discrete light beams.

5.2 Principle of Operation

The principle of operation of this detector can be seen from Figure 2. The scanner, operating in a plane at right angles to the conveyor, accepts only a pencil beam of light from the total light radiated by a tubular lamp. The acceptance beam is caused to sweep from top to bottom of the tubular lamp to form a plane of detection. When the leading edge of a parcel travelling on a conveyor reaches the detection plane it registers as an optical obscuration some time during the acceptance beam sweep. The parcel continues to register as a partial obscuration during each successive scan until the trailing edge of the parcel has moved past the detection plane.

5.3 Construction

The scanner consists of eight small lenses mounted symmetrically around the periphery of an opaque, light weight, hollow cylinder (see Figure 3). In the centre of the hollow cylinder is a static photo-electric sensor. A shroud prevents extraneous light reception via seven of the lenses, while light from the tubular lamp, as shown, is allowed to pass through the one remaining lens. The diameter of the cylinder is such that the distance between the illuminated lens and the photo-electric sensor is approximately the same as the lens focal length. It follows therefore that the photo sensor will only receive light accepted along a parallel beam, the diameter of which is controlled by an iris formed by the mount of the illuminated lens.

The sensor is a Light Activated Switch (LAS) incorporating an integral photo-diode and integrated circuit trigger element. The hollow cylinder with the eight lenses is caused to rotate around the stationary LAS such that each lens is illuminated in turn and as a result the tubular lamp is completely scanned by the light acceptance beam eight times during every revolution of the cylinder.

The sensor is constantly illuminated (except of course when a parcel is present) and must accept light over a 45° arc of detection. However, this condition can be satisfied only if the distance between the light source and the scanner is equal to the height of the light source. To avoid restricting the use of the detector, a second light source can be positioned behind an adjustable slit which directs a beam of light onto the sensor via reflection from a curved mirror (see Figure 3). Both the mirror and the associated light shield are pivoted about the same axis and can be adjusted independently of each other. The mirror is positioned so that its lower edge coincides with the upper limit of the required external scanning arc; the light shield is positioned so that the reflected rays are limited to 45° to the horizontal. The advantage of this arrangement is that the detector can be set up for a given scanning sector according to the site requirements, eg height of light source, and width of conveyor belt.

5.4 Optical System

The optical system was designed on the principle that the circle of confusion should lie within the silhouette of the smallest size object to be detected at the maximum range, ie adjacent to the light source. In practice this meant that the scanner was required to detect an object 10 mm high at a distance of 1 metre. This data enabled the permissible area of confusion and hence the maximum diameter of the accepted pencil beam of light to be defined. For optimum operation of the LAS it is desirable that the whole of the photo-sensitive area of approximately 1 mm^2 be illuminated. Using this and its relation to the permissible area of confusion defined by the above together with the fact that only 80% obscuration is necessary to provide the on/off discrimination for the electronics enables the iris diameter for the lenses to be fixed at 4 mm.

Allowing for losses in the optical system the available luminosity magnification is about 10 times.

The horizontal resolution (ie in the direction

of parcel movement) of the eight lens scanner driven at 600 r.p.m., giving 80 scans/sec, with a conveyor belt speed of 180 ft/min is 0.45 inches (11.43 mm).

5.5 Design of Photo-electric System

The principal problem in designing the photo-electric system was to obtain a source of light which would optimise the luminous and spectral sensitivites of the photo-electric sensor.

The threshold sensitivity of the LAS is determined by the choice of RC combination and provides a direct high level logic signal without supplementary buffering. The maximum switching frequency of the LAS is 20 KHz (equivalent to a pulse length of 50 uS) and is therefore adequate to detect the minimum pulse length produced by an object 10 mm high adjacent to the light source, ie 120 uS.

Tests showed that little loss in overall sensitivity occurred even if the lamp was run at half voltage. The probable reason for this is that the reduced luminosity of the lamp under these conditions is mainly at the high frequency end of the spectrum where the LAS is least sensitive. The sensor has low sensitivity to fluorescent tube illumination and is protected against direct sunlight by the narrowness of the scanning beam.

5.6 Other Applications

In its normal capacity the scanner is required to detect an object anywhere within the scanning field. If however the scanning rate were constant and the points of obscuration were timed relative to a datum angle, the simple on/off output could be used to measure irregularly shaped areas such as hides being processed by a manufacturer of leather goods.

6 CURRENT DEVELOPMENT IN LETTER MACHINERY

The next generation of letter processing machinery is being designed to take advantage of the greater reliability of modern electronics and there will be less dependence on mechanical components. A typical example is a re-design of the coding desk which incorporates closed circuit television as a convenient means of simplifying the letter presentation system, at the same time enabling the coding operators to be accommodated in a separate room away from the noise, heat and dust generated by the letter transport mechanism.

6.1 Early Experiments with CCTV

Early experiments were based on the principle that an address could be scanned whilst the letter was being transported at a relatively high constant velocity past a single camera. The technique was to use a short duration flash from triggered discharge tubes to illuminate the letter; the 'frozen' image projected on to the camera vidicon target was then stored on a track of a video disc recorder. As the recorder was equipped with an 8 track record and replay capability, up to eight images could be made available to coding operators.

The principle of 'freezing' addresses on fast moving items was fundamentally sound but there were many associated problems to be overcome before this technique could be put into practice. The most intractable problem was that of storing the letters in a fast moving transport system whilst they were being coded by operators and then keeping the coding information in station with the letters until they reached the printing position. Another difficulty was that there was no high speed code printer available to print code marks on moving letters.

6.2 Development of CCTV Presentation Systems for Field Trial

Although the 'frozen address' system was considered to be the ultimate objective, it was decided that an alternative scheme using stop/start techniques and known mechanical devices should be developed as an interim measure so that a CCTV system could be evaluated under sorting office conditions. The basic transport mechanism is shown in Figure 4. A simple inclined plane configuration is used incorporating a novel stop-start roller drive and a continuously moving belt down which letters travel with their short stamp edges leading. Each letter stops, in turn, in front of a camera and a printer. It was originally intended that each operator should have two such 'sidings' under his control and that the two cameras should be connected to two monitors (one above the other) at the operators console. The control circuitry would then switch the signals alternately from one siding to the other thereby achieving double presentation and increasing the coding rate. Unfortunately, the effect of two monitors is to introduce a very disturbing flicker sensation originating from the monitor which is not being viewed directly (the human eye is more susceptible to flicker when it occurs at the periphery of the field of vision than when it occurs along the axis). Furthermore, the distance between images is too large (although no more so than on the existing coding desk) and the capital cost of the system is too high. It is therefore proposed that the prototype equipment should be provided with additional circuits to generate the effect of double presentation by 'splitting' the image seen on one monitor (to which both cameras will be connected). Experimental work has shown that this can be effected relatively simply by synchronously sampling the video outputs from each camera in turn, one of which is supplied with vertical sync pulses delayed by half a field scan period (10 mSec) relative to the other (see Figure 5).

6.3 Other Factors

One of the requirements of the CCTV system is that the definition obtained should be sufficient for all types of addresses to be legible.

The video disc recorder and its associated equipment mentioned earlier use the standard 625 line system, but an 875 line system is being incorporated in the new equipment. This showed a marked improvement in the legibility of certain types of typewritten and printed addresses.

Another important advantage gained by the use of CCTV is the relative ease with which the form of image presentation can be changed. Various effects have been tried experimentally, including (1) alternate upper and lower image replacement, with and without blanking during the letter movement period, (2) "frame slip" which allows the letter's image to move either up or down the screen to reveal the next item to be coded, and (3) black/white reversal of the image. Some of these effects will be used in the prototype equipment for evaluation purposes.

7 FUTURE DEVELOPMENT

Solid state light sources are being used in increasing numbers in most of the equipment. Filament lamps are being replaced by visible light emitting diodes (VLED's) on control panels and on printed circuit boards, and gallium arsenide sources will be used in most of the photo-beam applications described earlier. The infra-red radiation from the latter devices matches the peak sensitivity of the photo-transistors used as detectors, resulting in reduction of interference from ambient lighting.

Display units, comprising a matrix of small VLED's in dot or bar form, will be used extensively in display and control panels for machines and testers. A simplified version of a refresher-training unit for coding desk keyboard operators has been constructed using a small computer in conjunction with these devices instead of the more complex cathode ray tube visual display units. The advantage of this system lies in the large cost savings made in the data presentation, which could result in its installation in every Mechanised Letter Office.

Liquid crystal (nematic) displays, with their lower power requirements and larger character size, will no doubt displace the LED versions when the life expectancy of the former is improved.

Optical couplers, consisting of an LED-photo transistor combination in the same package, are being incorporated in many control circuits currently being designed. They are used to link two parts of a circuit without the need for a direct connection, and since the coupled circuits are completely isolated from each other, they offer an attractive way of eliminating earth loops in situations where machines have to be inter-connected for data transmission purposes.

Future parcel offices are likely to have a greater degree of automatic control. To provide a smooth parcel flow sensing devices which will indicate rate of mail movement, instead of the on/off type of detectors currently in use, are required. Optical scanners incorporating lasers are now commercially available and are being used in warehouses to read storage addresses. Similar equipment could be installed on parcel sorting machines to permit automatic machine reading of addresses for bulk postings of parcels, provided that suitably coded labels are previously attached.

Many advances can be expected in the applications of closed circuit television. Solid state replacements for vidicons are becoming more readily available and their resolution is being continually improved. If the introduction of these devices results in a cheaper form of TV camera, a system could be developed in which all sides of a parcel could be simultaneously displayed to the sorter, thus eliminating the need for manual facing during the sorting process. The simplified cameras would also result in more compact and robust units, and a digital output signal may prove to be superior to the analogue version in certain applications.

8 ACKNOWLEDGEMENTS

Acknowledgement is made to the Director Engineering, Postal Mechanisation and Buildings Department, for permission to publish this paper. Acknowledgement is also due to Mr D G Hunt, Mr H R Henley, and Mr Graham B Kent (Consultant Engineer), the inventors of the parcel detector, and those who assisted in the preparation of this paper.

9 REFERENCES

de Jong, N C C. Progress in Postal Engineering. Part 2 - Packets and Parcels. POEEJ Vol 63 p131 Oct 1970.

de Jong, N C C. Progress in Postal Engineering. Part 3 - Letter Mail. POEEJ Vol 63 p203 Jan 1971.

Gubbay, D M, and Wiles J R. Re-orientation of Moving Letters. I.Mech.E. British Postal Engineering Conference, May 1970, Paper No 39.

Copping, G P. Automatic Letter Facing. I.Mech.E. British Postal Engineering Conference, May 1970, Paper No 17.

Wicken, C S. Control Systems for Automatic Letter Facing Machines. I.Mech.E. British Postal Engineering Conference, May 1970, Paper No 18.

Adams, G F. Letter Facing Machine: Development and Design of Production Machine. I.Mech E. British Postal Engineering Conference, May 1970, Paper No 41.

Knowles, S C. Development of a Letter Coding Desk. I.Mech.E. British Postal Engineering Conference, May 1970, Paper No 22.

Beardmore, A F. Storage Problems in the Parcel Post. I.Mech.E. British Postal Engineering Conference, May 1970, Paper No 16.

Castellano, E J. Tilted Belt Parcel-Sorting Machine. I.Mech.E. British Postal Engineering Conference, May 1970, Paper No 6.

Hills, E G., Wicken, C S., Gubbay, D M and Worwood, G S. The Stacker Link in the Facer-Canceller Table Application. Letter Machinery Branch Report No 4. Nov 1970.

Bennett, H A J., and Spence K L. A Computer Based Coding Desk Operator Training System. Letter Machinery Branch Report No 11. June 1971.

Maximum width of conveyor: 5 ft

Belt speed: 180 ft/min

Maximum vertical variation in
belt position in the vertical
plane: 0 to 0.25 inch.

(including dirt on belt surface).

Vertical resolution required: 0.375 inch, from 0 to 6 inch above belt.

                                    0.75 inch, from 6 to 12 inch above belt.

                                    1.25 inch, from 12 to 17 inch above belt.

                                    (Better preferred if economically possible).

Accuracy with which the
detector must determine the
horizontal position of the
parcel along the belt: 0.4 inch band for all conditions.

Approximate minimum
spacing between parcels: 0.5 inch at 180 ft/min.

Letter Tray Showing Code Reflectors

Fig.1

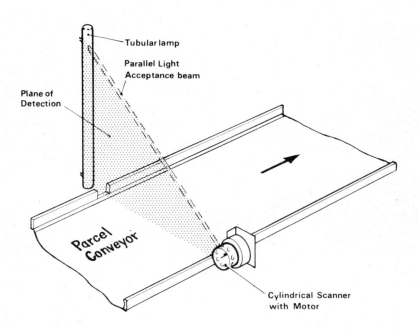

Basic Layout of the Continuous Scanner

Fig.2

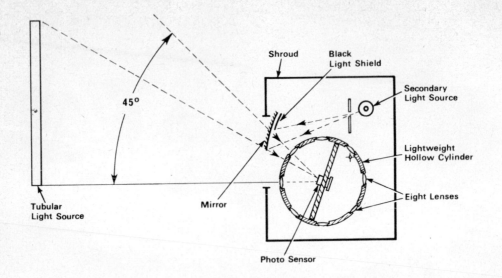

Continuous Scanner Hollow Cylinder Arrangment and Arc Adjustments

Fig.3

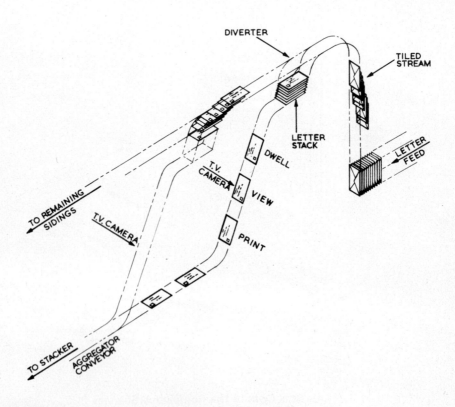

FIGURE 4. C.C.T.V./L.T.S. LETTER PATH DIAGRAM

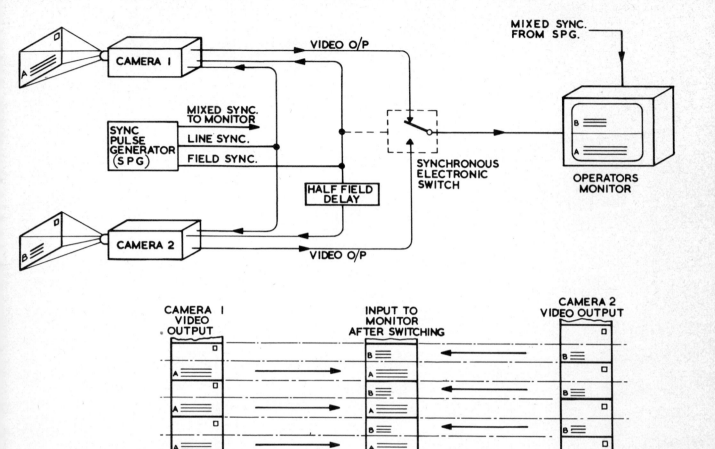

FIGURE 5. CCTV "SPLIT-SCREEN" EFFECT FOR LETTER PRESENTATION

C124/74

VIDEO STORAGE TECHNIQUE FOR A LETTER CODING SYSTEM

WILLIAM JAMES ROWLEY CLARK
Engineering Manager, Marconi Elliott Avionic Systems Limited,
Electro Optical Systems Division, Paycock Road, Basildon, Essex
The Ms. of this paper was received at the Institute on 10th December 1973 and accepted for publication on 31st January 1974. 33

SYNOPSIS Automated postal sorting machinery based on the addition of a machine readable dot code to the envelope has now become widely accepted by postal authorities. As an alternative to interrupting the letter flow whilst an operator types the sorting code, a short pulse illuminator combined with a C.C.T.V. camera can be used to effectively immobilise an image of the letter as it passes the viewing station. This transient image is then stored on a Silicon Storage Vidicon and read out non-destructively to present a repetitive stationary image on a standard television display monitor.

1. INTRODUCTION

Automated postal sorting machinery based on the addition of a simple machine readable address code to the envelope has now become widely accepted by postal authorities. In most present systems, each envelope is directly viewed by an operator who translates the address into a dot code by means of a manual keyboard. At each operator station letters must be stopped and presented for viewing before being inserted into the conveyor stream for delivery to the automatic sorting machine. In the large sorting offices there may be as many as 150 of these operator stations, so that a significant proportion of the cost of an automatic sorting installation is attributed to the mechanical handling machinery at the addressing stations. If, however, means can be found for the operators to read the address whilst the envelopes are in motion on the conveyor, significant economies can be made in mechanical handling costs. To achieve this, two fundamental requirements must be satisfied; firstly an effective immobilisation of the image at the viewing station, and secondly a means of presenting this stationary image to the operator for as long as he requires it to perform the coding function. The use of a television camera combined with a pulsed light source at the viewing station provides the required image immobilisation whilst a video storage tube permits continuous viewing of the image for as long as the operator requires. Design features of the equipment required to immobilise the image and to present a stored image to the operator will be discussed in the sections which follow.

2. IMAGE IMMOBILISATION

A letter flow rate of 3 metres/sec. can be considered typical of a modern conveyor. If a television camera is arranged so that a 22 cm long envelope just fills the scan then each letter will cross the field of view in 73.3 ms. This will, of course, give rise to a quite unacceptable image blurring if directly viewed and so a means of immobilising the image is a primary requirement of the television system. This can be quite simply achieved by utilising the reciprocal relationship exhibited by camera tubes between intensity of illumination and exposure duration. In the case of the vidicon camera tube the charge established on the photosensitive target layer bears a constant relationship to the intensity time product down to exposure durations of about 1 ms. For shorter pulse lengths, the effect of charge build-up lag starts to become significant but nevertheless providing the intensity is sufficiently high very short exposure times can be employed. By ensuring that the image movement is negligibly small during the exposure period, an effective means of eliminating the effects of image motion is available.

For a balanced resolution system the horizontal resolution (R) will be given by the expression

$$R = AK (N-N_b) \qquad (1)$$

Where N = the number of scanning lines

N_b = the total number of lines lost in the frame blanking intervals

K = the Kell factor = 0.7

A = aspect ratio = $\frac{4}{3}$

For the standard 625 line system

$R = \frac{4}{3} \times 0.7 (625-40) \approx 540$ lines/picture width.

Referred to the plane of the envelope being viewed, this corresponds to a resolution of 0.4 mm when a 22 cm envelope just fills the field of view.

If the letter is imaged onto the camera tube photosensitive layer for a time which is small compared to the time for the letter to move by one resolution element then the image will be effectively immobilised and no significant loss of image quality will occur due to image motion.

The relationship between exposure time t, object velocity V, and allowable distance travelled will be simply.

$$t = \frac{d}{V} \qquad (2)$$

Thus, if we judge that 0.1 mm of object movement will be unnoticeably small the required exposure duration is 33.3 µs.

In principle this short exposure could be achieved either by a shutter or, more conveniently, by using a pulsed illuminator. Xenon flash tubes can operate quite efficiently down to pulse lengths of about 5µs and since high intensity levels are readily obtainable they are found to be the most suitable for this application.

Quite clearly, the whole feasibility of the pulse illumination concept hinges on establishing that an adequately large charge pattern can be stored on the camera tube target during the exposure period with currently available flash tubes. If, initially, we assume a strictly reciprocal relationship between the level of illumination and the time of exposure, then the necessary illumination can be deduced from the steady state illumination characteristic of the camera tube. Based, for example, on data given for the P 842 vidicon, a continuous illumination at 2854°K of 108 Lux produces a peak signal current of 0.3μA at prescribed electrode potentials and can be considered typical of the conditions to produce good imaging quality. Under steady state illumination conditions, the charge continues to build up during the frame period of 1/25 sec., part being removed on each one of a pair of interlaced field scans and a residual amount being read-off on subsequent frames. As a first approximation the total light contributing to the charge pattern will be 108/25 = 4.32 Lux-seconds.

If however, the vidicon target is illuminated once per frame with a pulse of length t, each flash will be required to produce an illumination of $\frac{4.32}{t}$ Lux if the peak signal current is to be the same as with continuous illumination. To be strictly accurate, this expression should be modified to take account of the fact that whereas the steady state illumination figures are quoted at a colour temperature of 2854°K, the Xenon flash tube spectral response does not follow the black body curve over the spectral band of the camera tube. However, an analysis of the effect of the spectral mismatch shows the correction factor to be 0.956 which is so close to unity as to be quite negligible for all practical purposes.

A more serious difficulty arises through our initial assumption that a strictly reciprocal relationship existed between illumination and pulse time. In fact, it is known that reciprocity failure occurs when the pulse times are short, due to an effect which is known as charge build-up lag. The effect of build-up lag on reciprocity is believed to start to become significant at pulse lengths of about 1 ms and is therefore clearly a factor of importance at the shorter pulse length needed to ensure effective immobilisation of the image. In the absence of any reliable data on the build-up lag characteristic, a correction factor has been established experimentally at a pulse length of 31μs. The experimental results, which were based on a direct comparison of a Xenon flash tube and a calibrated tungsten lamp, have suggested that a 20 fold increase in light level is necessary over that which might be predicted from a strictly reciprocal relationship. Thus to obtain a charge equivalent to the steady state case requires a pulsed illumination of $(4.32/33 \times 10^{-6}) \times 20 = 0.26 \times 10^7$ Lux. An EDN 10 Xenon strobe flash (with a reflector) has a rated axial intensity of 2×10^6 candelas at the appropriate pulse length and at a distance of 0.43 metres from the letter, this corresponds to an illumination of 1.08×10^7 Lux which exceeds the requirement by a factor of 4 : 1. Results achieved with an experimental system employing this flash tube with a standard P842 vidicon have confirmed the adequacy of the illumination margin and the effectivness of this means of providing image immobilisation.

Figure 1 shows the way in which a Xenon flash tube illuminator can be combined with the television camera to form a complete viewing system.

The presence of a letter is first sensed as it enters the viewing station by means of a light beam and a photo detector. To ensure that the whole of one image is read-out on the same scanning raster, the flash tube must be synchronised so as to occur only during the field retrace interval on either the odd or even field of the television frame. This can be readily achieved by arranging for the first or second field drive pulse following interruption of the beam to be used for timing the flash tube trigger. The random time intervals between the sensor pulses and the first field pulses can be as much as 20 ms, consequently the positional uncertainty of the letter image at the object plane will be ± 3 cm. However, the time delay between sensor pulse and the lamp trigger can be used to give a direct measure of positional error and so offers the posibility of a dynamic raster shift compensation on the camera to eliminate the positional error on the displayed image automatically.

The maximum capacity of the television system for viewing a stream of letters passing along the conveyor will be set by the frame rate to 25 letters/second (or 90,000 letters/hour). However, it is a characteristic of a Vidicon camera tube that a single frame scan does not completely erase the stored charge on the photosensitive layer and this image 'lag' effect would result in a cross talk signal from one frame to the next. A preferred method of operation is to utilise one frame for viewing purposes whilst reserving one whole frame for residual image erasure. A third frame is required for letter sensing and charge stabilization. Although wasteful of capacity, this still offers the potential for recording up to 30,000 letters/hour from a single camera station which is likely to be adequate for even the largest mechanical handling systems.

Figure 2 shows a photograph of the output waveform of the television camera immediately following firing the flash tube. It consists of the odd and even field containing the wanted video signal, followed by subsequent fields during which the residual charge is read out at decreasing amplitude. It is of particular interest to note the way in which the video signal builds up during the first field after the flash exposure and only achieves full amplitude in the second field. This signal build-up is believed to be due to an effect which has been called 'photo-conductive' storage. Some of the current carriers generated within the target layer fall into thermal traps which have a relativly long time-constant so that carriers which are subsequently released only become available for conduction some time after the incident light has ceased. The number of carriers caught in these trapping centres is quite small at light levels at which a vidicon tube is usually operated and the effects of photoconductive storage can usually be ignored. In this application, however due to the extremely high light level, the effect is most marked and results in the severe shading shown in the first field waveform. Fortunately the effect can be overcome by the simple expedient of gating off the scanning beam for the first field following exposure and only reading out the video signal on the second and third scans. Figure 3 illustrates this preferred sequence which overcomes the effect of photoconductive storage. In practice, the errase fields would most probably be blanked out to provide a convenient time interval for signal switching in the subsequent signal routing equipment.

It will be appreciated that the television camera channel used to produce the train of immobilised letter images can be a perfectly standard product in all other essential respects. There may be some advantage in optimising certain of the signal processing circuits to maximise the reproduced picture quality, bearing in mind the rapidly changing signal content from frame to frame; but such changes are relatively insignificant.

The very fact that one T.V. camera can handle all the video data required for a large sorting install-ation demands that an exceptionally high standard of reliability must be established in this one item of equipment. In practice, although modern, high grade, C.C.T.V. cameras are highly reliable, the camera tube itself cannot be guaranteed beyond about 1000 hours operation and many tubes will have failed within 5000 hours. In this situation there may be a real operational advantage in duplicating the camera station to provide a ready to use spare and thus reduce to acceptable proportions the probabil-ity of operational down time.

3. PRESENTATION OF THE SIGNALS TO AN OPERATOR

The train of televised images appearing at the out-put of the camera channel, although containing all the required video information, cannot, of course, be directly viewed by an operator.

Each individual frame of information must first be stored and then presented to the operator, on demand, for as long as is necessary for him to carry out the addressing sequence. A viewing time of 2 seconds might be considered typical and this corresponds to 50 complete television frames if a standard television scanning rate is chosen for the read-out process.

All fields in the sequence should be nominally identical and this implies that the video store should have a non-destructive read-out characteristic for at least 2 seconds and ideally for a significant-ly longer period to allow for operator interruptions. The store must then be capable of speedy erasure so as to again become available to accept a new frame of picture information from the incoming letter stream.

In principle, a number of different types of video storage schemes might be considered for the appli-cation including digital storage, magnetic tape, or disc machines. Since, however, there must be at least as many storage devices as operators, the chosen method must be both inexpensive and reliable. With digital methods, for example, a storage capacity of about 10^6 bits/frame would be necessary at each operator station and estimates suggest that this would involve equipment costs at least 10 : 1 higher than alternative analogue storage methods. Of the analogue methods, either multi-track Magnetic Disc Stores or Silicon Storage Vidicons might be consid-ered practical solutions at acceptably low cost. For a small number of operator stations, the Silicon Storage Vidicon is undoubtedly the cheaper of the alternatives with the advantage favouring the magnetic disc stores as the number of stations is increased. However, bearing in mind the substant-ially continuous operation required from the storage bank, operating life and reliability are considered to be of primary importance in making the final choice and here the advantage undoubtly lies with the Silicon Storage Vidicon. The fact that each store is independent of others in the banks instead of partially interdependent as in the multitrack disc store, is particularly important in assessing the relative merits of the alternative analogue solutions.

On balance, therefore, the Silicon Storage Vidicon appears the preferred choise as the basis for the design of a system complex. Because it is a derivative of well established family of vidicon camera tubes, it can make use of the relatively inexpensive scanning and focusing components devel-oped for the C.C.T.V. market and the associated

electronic circuits can be readily derived from those available in proprietary cameras.

4. THE SILICON STORAGE VIDICON

The Silicon Storage-Vidicon is a single gun vacuum tube closely resembling the Silicon Vidicon but with a storage element replacing the normal photosensitive target. Silicon dioxide 'islands' are formed on the surface of a disc of silicon to act as the charge storage elements and these are scanned by an electron beam which performs the reading, writing or erasing functions as determined by the potenti-al applied to the target electrode.

The principles of operation are illustrated diagrammatically in Figure 4. To erase any previously stored charge the target electrode is made some 20 volts positive with respect to the cathode potential whilst the storage surface is scanned line by line with an unmodulated electron beam. In this mode the secondary emission co-efficient of the target is less than unity so that the potential of the 'islands' is brought to cathode potential whilst the target remains at +20 volts. (See Figure 4 (a)).

5. WRITE

To 'write' a charge pattern, the target is raised to + 150V, the silicon dioxide islands remaining 20V less positive at + 130V. The scanning beam is modulated by the video signal. Electrons landing on the islands will generate secondary electrons which will be absorbed by the more positive areas of the target between the 'islands'. In this way the 'islands' become more positive where the beam mod-ulation is highest, so causing a charge pattern corresponding to the beam modulation. (See Figure 4 (b)).

6 READ

The charge pattern is read out be reducing the target potential to + 9V which leaves all the 'islands' more negative than the cathode (Figure 4 (c)). The unmodulated beam is now scanned across the target and is repelled by the more negative 'islands'. The number of electrons reaching the more positive target, and constituting the signal current, is controlled by the negative potential of the 'islands' in a similar manner to the action of the control grid of a triode valve. The residue of the beam is collected by an adjacent mesh.

Since virtually no electrons land on the 'islands,' the stored charge information is unchanged by the reading process so that the read-out is substantially non-destructive. In practice the presence of stray ions cause a deterioration of the charge but only after many minutes of repeated scanning.

7. SYSTEM CONSIDERATIONS

The television viewing system described offers considerable flexibility for system design and, clearly, the exact form adopted will be dependent on the overall configuration of the sorting office. However, for illustrative purposes, a general scheme is shown in Figure 5. In this embodiment, letters are de-stacked whenever a space is available on the conveyor loop and a video store is simultaneously available. Letters which have been actioned by the operator are removed from the conveyor loop as they pass the divertor and proceed to the address printing equipment for coding prior to being automatically sorted.

Figure 6 is a schematic representation of the television signal switching circuits. In this case 'n' video stores are provided to service the requirements of 'm' operator stations and electronic switching of the video signals is provided at the input and output of the stores. A tell-back signal is provided from each store to indicate whether it is in an 'erased' state or if it is storing a field of video data. As each new field of video data is generated by the camera, the input switcher interrogates each store in turn and connects the video signal to the first available store found in the erased state.

The input switcher sucessively interrogates from 1 to 'n' and back to 1 again, ignoring the stores which are in a reading mode and therefore loading the stores in a sequence which is random in relation to the original letter stream. Similarly, on the output side, an operator demand for a fresh letter, initiates erasure in the store to which he was connected and causes the output switcher to scan down the bank of stores to identify the first unit whose tell-back indicates the store is filled. Once connected, the read-out continues until the operator makes a new demand which repeats the output switching sequences.

In a system of this type it is, of course, necessary to retain the correct sequence between the original letter stream and the sequence of address codes fed to the coding dot printer at the down stream end of the conveyor. To achieve this, despite the random loading of the video stores, each new frame of video information generated by the television camera has a sequence code added in the form of a binary pulse waveform impressed on the television waveform during the field blanking interval. This sequence code, which is, of course, stored with the video information, is automatically read out at the operator's console and prefixed to the postal code typed in by the operator. This simple frame marking system allows the data handling equipment to order the 'address code' data from the keyboards and thus regenerate the original letter input sequence at the code dot printer station.

8. CONCLUSIONS

A television camera combined with a short pulse illuminator provides an effective means of immobilising the image of a letter moving on a conveyor whilst the Silicon Storage Vidicon provides a satisfactory means of presenting a stationary, repetitive picture to an operator at a coding desk. Results achieved on an experimental model have demonstrated the esential feasibility of the system described and the adequacy of the illumination levels achieved from a Xenon flash tube.

The television method should offer a means of reducing the overall system cost and of improving reliability due to a simplification of the mechanical handling equipment. It also removes the need for the coding desk operators to be physically located near the mechanical handling machinary, permitting an improvement in their working environment and offering greater freedom in space utilisation.

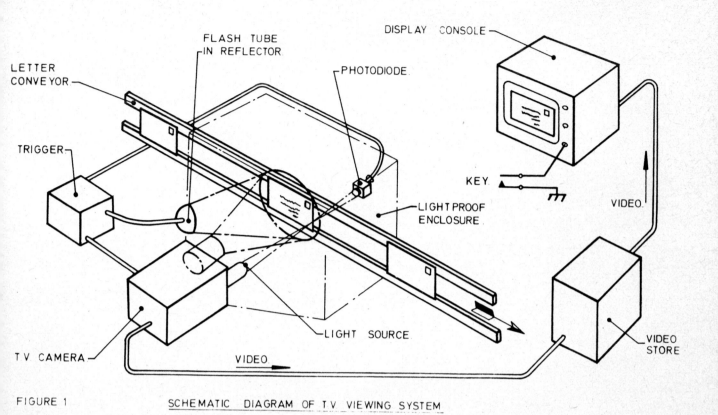

FLASH TUBE
IN REFLECTOR.

DISPLAY CONSOLE

PHOTODIODE.

LETTER
CONVEYOR.

TRIGGER

LIGHT PROOF
ENCLOSURE.

KEY.

VIDEO.

LIGHT SOURCE.

T.V. CAMERA

VIDEO
STORE

VIDEO.

FIGURE 1   SCHEMATIC DIAGRAM OF T.V. VIEWING SYSTEM

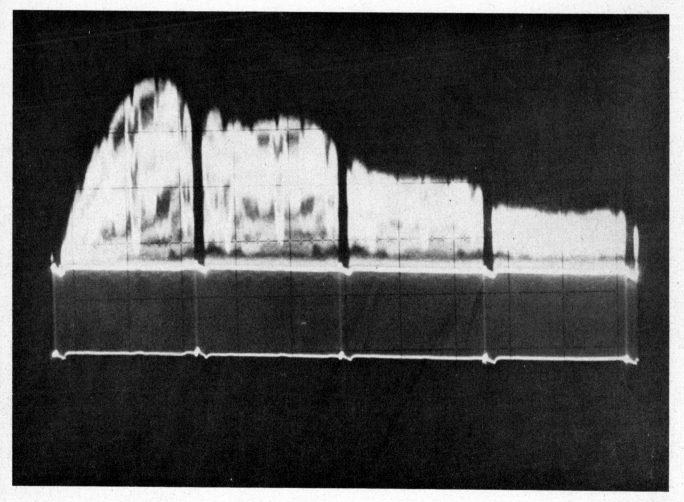

Fig. 2. Television Signal Waveform.

23

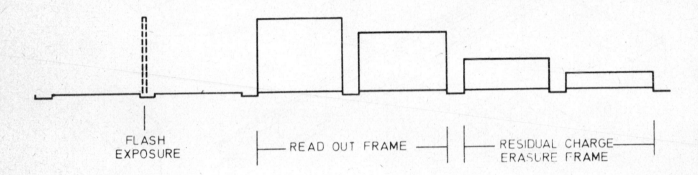

FIGURE 3

TELEVISION   CAMERA   READ-OUT· SEQUENCE

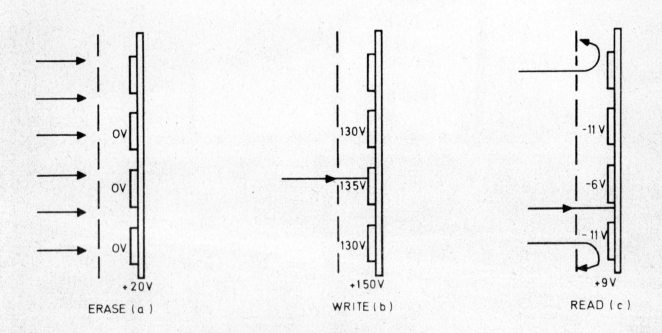

OPERATION OF SILICON STORAGE VIDICON

FIGURE 4

24

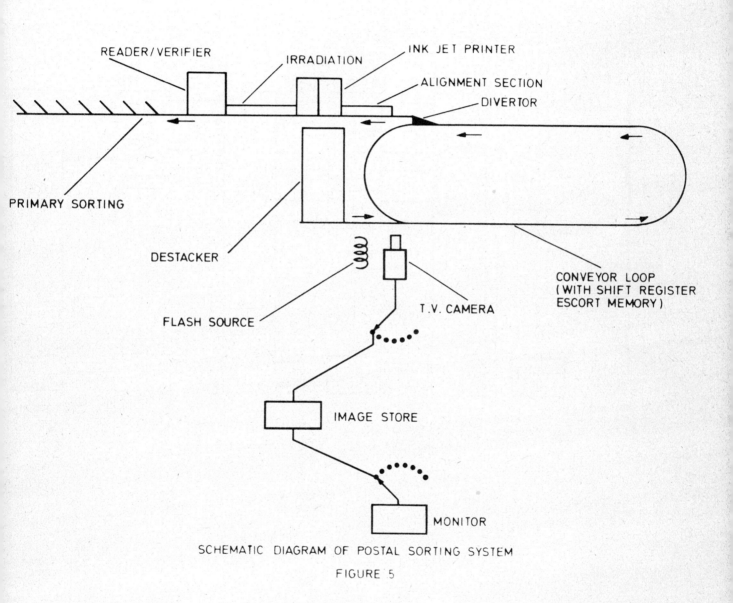

SCHEMATIC DIAGRAM OF POSTAL SORTING SYSTEM

FIGURE 5

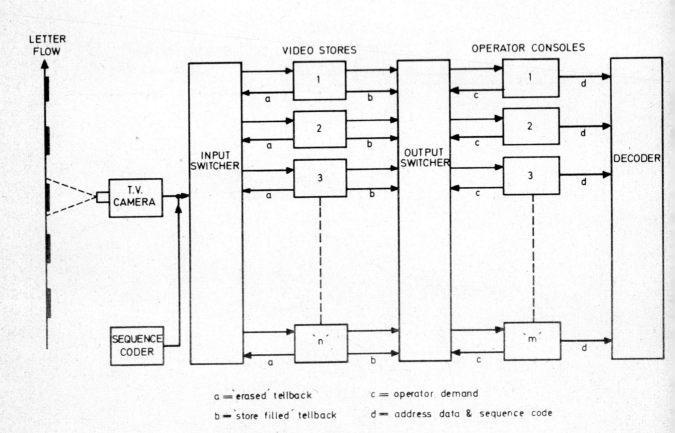

a = 'erased' tellback   c = operator demand

b = 'store filled' tellback   d = address data & sequence code

ELECTRONIC SWITCHING ARRANGEMENT

FIGURE 6

C125/74

AUTOMATIC OPTICAL SORTING MACHINES

WILLIAM STOREY MAUGHAN, B Sc (Hons), Ph D
Research Manager, Gunson's Sortex Limited,
Fairfield Road, London E3
The Ms. of this paper was received at the Institution on 24th January 1974 and accepted for publication on 18th February 1974. 23

SYNOPSIS   Introduction to automatic sorting machines based on 'colour' measurement. Use of a Particle Spectrophotometer to determine surface reflectivity characteristics. The choice between monochromatic and bichromatic measurement. Selection of suitable lamps, filters and detectors. Illumination requirements. The matched background concept and the need for scanning systems. The mechanical design of optical inspection chambers. Mechanical details of feeding and separation devices. Details of the signal processing equipment. Size and/or shape measurement using electro-optical devices.

1. INTRODUCTION

The sorting of particulate matter into accept and reject categories is one of the oldest of man's occupations. The food and mineral industries are probably two of the areas in which such sorting has been practised for the longest period of time and it is the purpose of this paper to describe some of the techniques and machinery that have been introduced into these industries in order to automate the sorting process.

Hand sorting has been and still is widely practised and as this depends on a visual evaluation of the particles it is not surprising that the vast majority of automatic sorting machines have been based on some form of optical assessment of the particles. Of course, other types of measurement (e.g. X-ray, magnetic, radioactive etc.) are possible and indeed in some cases are essential, but the present paper is restricted to optical measurements.

In order to replace the hand sorter a machine has obviously got to duplicate some of the functions of the human eye, brain and hand. Machine sorting of this type is often referred to as 'Electronic Sorting' as it was the advent of electronics that made possible the duplication of the eye and brain functions.

Significant experimentation with automatic sorting equipment for both food and mineral applications started immediately after World War 2. Colour sorting machines gained early acceptance in the food industry (1948 onwards), their introduction into the mineral industry started in the early 1960's.

An increasing demand for automatic sorting machines is ocurring due to the escalating costs of hand sorting coupled with the higher quality requirements being imposed on food producers and an increasing realization of the importance of sorting in the reduction of health hazards arising from contaminated food.

An optical sorting machine will, in general, consist of three basic parts, (Figure 1)

(a) A feeding system, to feed the particles in a controlled manner to the inspection system (see section 8).

(b) An inspection system to measure the optical reflectivity of the particle and classify the particle as accept or reject (see section 7).

(c) A separation system capable of physically separating those particles classified as accept from those classified as reject (see section 9).

The size, cost and complexity of sorting machines varies depending on the size range of particles to be handled, the type of optical measurement being made and the throughput requirement. Although many machines are employed in sorting particles as small as mustard seeds, probably the smallest particles that are sorted on a large commercial scale are rice and 0.5mm diamonds, whilst the largest are 150mm rocks (limestone, magnesite etc) The throughput of a single channel of rice sorting may be 50kg/hr. whilst a single channel rock sorter might handle 50 t/hr. Optical measurement can be made by using a simple two direction monochromatic determination or a three directional bichromatic scanning system (see sections 3 and 6) Capital costs of the machine alone can vary from £2,000 to £20,000 each.

The following list, which is by no means complete, will serve to indicate the range of products at present handled by automatic optical sorting machines:

Food;		Minerals;	
	Coffee		Diamonds
	Rice		Limestone
	Peanuts		Rock Salt
	Beans		Marble
	Maize		Gypsum
	Almonds		Chalk
	Walnuts		Magnesite
	Frozen Peas		Talc.
	Frozen Potato Chips		
	Whole Peeled Potatoes		
	Carrots		
	Cherries		

2. PARTICLE SPECTROPHOTOMETER

To successfully separate one type of particle from another by optical sorting, it is obviously necessary to know something of their surface reflectivity characteristics. Published reflectivity figures for different materials usually only apply to carefully prepared surfaces of the material measured under precisely defined optical conditions. In a practical sorting machine, however, we are dealing with a naturally ocurring surface viewed under compromise light conditions. The Sortex Particle Spectrophotometer is an instrument which has been specifically developed to make reflectivity measurements on natural particles under conditions which closely resemble those found in sorting machines.

3. MONOCHROMATIC AND BICHROMATIC SORTING

3.1. Monochromatic sorting

The simplest form of optical sorting is the Monochromatic system. As the name implies, this is sorting based on a measurement of reflectivity over a single band of wavelengths. Suppose the results of the measurements on a number of particles using the Particle Spectrophotometer are as shown in figure 2. Each particle has a spectral response curve which lies within one of the two areas shown. Clearly if an unambiguous separation is to be made between particles in one area and those in the other then the measurement should be confined to a wavelength where the two areas do not overlap. In the case depicted in figure 2, the waveband from 500nm to 650nm would be utilised.

3.2. Bichromatic sorting

Unfortunately, it is not always possible to find a section of the spectrum where the accept and reject areas are clearly separated and in these cases it is necessary to resort to a more complicated procedure which involves measurement in two different wavelength bands. This is known as Bichromatic sorting. Figure 3 shows two families of spectral response curves, one corresponding to the accept particles and the other to the rejects. Clearly there is no wavelength band where there is a clear separation of the two families. However, if measurements are taken in the wavelength bands A and B and then the ratio A/B is calculated it will be noted that A/B is greater than one for all accept particles but less than one for the rejects. The ratio of the two measurements can therefore be used to give an unambiguous separation of the two particle types.
Some practical separations require both monochromatic and bichromatic decisions for the successful removal of all defects and therefore automatic sorting machines are available which are capable of making both types of measurement simultaneously. The technique is clearly extendable to trichromatic measurement but no practical machines have been introduced with this feature.

4. LAMPS, FILTERS AND DETECTORS

At any given wavelength the electrical signal from the detector will depend on four main factors:

(a) The light source
(b) The particle reflectivity and size
(c) Any optical filters fitted to the detector
(d) The detector

Having decided, by the methods outlined in sections 2 and 3 above, the wavelength band or bands over which measurements should be made, factors (a), (b) and (d) have to be considered with the object of ensuring the maximum possible signal/noise ratio from the detector in the required wavelength band(s) and minimum signals in all other wavelength bands.

4.1. Lamps

Two basic types of lamp are used as light sources in optical sorting machines.

(1) The fluorescent tube, and (2) the incandescent filament bulb.

Each of these types has certain advantages and disadvantages and the principle of these are listed below:

Fluorescent Tube:

Advantages -	Cool operation Long life Diffuse source
Disadvantages -	Low on red emission Requires AC power supply (High frequency inverters)

Incandescent Filament:

Advantages -	Wide Specular emission DC operation
Disadvantages -	Excessive heat produced Point source

In general, the use of fluorescent tubes is favoured except in cases where a deep red or near infra-red measurement is called for.

4.2. Filters

A wide range of optical filters exist, both of the colour glass and interference types and it is therefore usually a simple matter to select suitable filters to isolate the required band(s) of the spectrum. Four basic classes of filters are utilised:

(1) Low Pass   - transmitting only below a certain wavelength
(2) High Pass - transmitting only above a certain wavelength
(3) Broadband - transmitting only within a broad band (50nm) of wavelength
(4) Narrowband - transmitting only within a narrow band (10nm) of wavelength

4.3. Detectors

Until recently the photomultiplier tube, in all its various forms, offered the best performance as a detector of visible radiation. Its chief advantages are its good signal/noise ratio, enabling it to detect low light levels and its satisfactory blue end response. Its disadvantages, however, include its mechanical construction, need for high voltage supplies, and lack of deep red and near infra-red response.
The development of the solid state photodiode has, however, almost reached the stage

where it is able to match the photomultiplier in most cases except those involving measurements at the blue end of the spectrum. The solid state detector has the advantages that it does not require a high voltage supply, it is compatible with modern solid state electronic circuitry, it is cheaper, it is mechanically robust and has an indefinate life. Unfortunately, these sensors are low on blue end response and therefore, although modern sorting machines employ solid state sensors wherever possible, the photomultiplier has still to be used where a critical blue measurement is required.

5. ILLUMINATION REQUIREMENTS

When dealing with irregular shaped particles, uniform diffuse illumination is necessary to minimise the occurrence of highlights and shadows which could obviously detract from the measurement of true surface reflectivity. Ideally, at the point of measurement, the particle should be surrounded by a spherical surface of uniform brightness. In practice however, a number of factors preclude the attainment of this ideal. Firstly, the particle has to enter and leave the inspection chamber in free fall and this requires the provision of entry and exit ports in the top and bottom of the chamber. Secondly, the inspection optics, backgrounds etc. result in areas of different brightness to the main chamber wall and thirdly, the use of light sources of finite size leads to non-uniform illumination.

A further problem that can arise even with the perfect diffuse illumination sphere is that of specular reflection. If a particle with a perfect diffuse reflective surface is placed in a diffuse illumination sphere then its true colour will be observed. However, if the particle surface is not diffuse, then specular reflection occurs, giving highlights which do not exhibit the true colour of the surface. Clearly the highlights can give incorrect information to the optical system and hence result in the wrong classification of a particle.

The practical optical chamber usually contains either a number of fluorescent tubes arranged to give as uniform a distribution of light as possible or a number of incandescent filament bulbs with screens positioned to stop any direct radiation from the filaments to the particle (see section 7).

6. MATCHED BACKGROUND AND SCANNING SYSTEMS

Having arranged for the particle to be illuminated in the optimum manner, the next step is to consider the best optical arrangements for obtaining the necessary reflection data.

6.1. Matched Background System

The simplest form of inspection system is the matched background illustrated in figure 4. An illuminated background is chosen which, from the viewing direction appears to have a brightness between those of the accept and reject particles.

This system has a number of advantages, the principle one of which is that it is size independant. As an example, consider the case where it is required to eject dark particles

from acceptable light ones. If a matched background is used, then whenever a dark particle passes through the viewing zone a decrease in signal will result, but a light particle will cause an increase in signal, thus an unambiguous decision can be made by the electronics. If, on the other hand, a background lighter than all particles were to be used, then all particles would give a decrease in signal and in particular the small dark particles would give identical signals to large white ones and hence these two could not be separated.

It should be noted that the viewing area is in the form of a slit, the width of which is chosen to allow for the scatter in the trajectories of particles and for the size range of particles being sorted. The height of the slit is kept to a minimum, consistent with sufficient signal, in order to give maximum resolution and for accurate timing of the delay between inspection and separation. This arrangement gives good vertical resolution but inferior horizontal resolution.

6.2. Scanning Systems

Where small areas of discolouration are to be detected it may be that the slit system gives insufficient horizontal resolution and it is then necessary to go to some form of scanning system.

The first type of scanning system involves the use of a rotating aperture disc (figure 5a). This system results in only a small area of the slit being viewed at a given time, but this area scans across the total width of the slit in a similar way to the line or horizontal scan of a television tube. The vertical or frame scan is, of course, provided by the motion of the particle through the viewing zone and thus the whole of the particle is scanned by the small area.

In the second type of scanning system (figure 5b) the viewing zone is split up into a number of areas by the utilisation of a solid state detector which consists of a number of separate elements (a photodiode linear array). The information from these elemental detectors is then scanned electronically.

As well as giving a higher resolution the scanning system can also be utilised to give information on both the area of the particle and the area of blemish or discolouration thus allowing the sorting of particles on the basis of their percentage area of blemish.

7. DESIGN OF OPTICAL CHAMBERS

The ideal type of optical inspection chamber has already been described (section 5) and a number of the practical problems considered. In designing a practical inspection chamber the following requirements have to be considered.

(a) The illumination of the particle at the inspection point must be as uniform as possible.
(b) Lamps must be mounted so that they can be easily replaced.
(c) Direct light from non-diffuse light sources must not fall on the particle at the inspection point.
(d) The particle must not be allowed to come into contact with any part of the chamber, particular care being necessary with hot surfaces (e.g. incandescent bulbs).
(e) Backgrounds must be provided which can be readily changed and which can preferably be

adjusted by mechanical movement to give different brightnesses.

(f) Lens tubes require careful design and positioning to avoid shadowing or highlight effects on the particle at the inspection point.

(g) All critical optical surfaces (lenses, backgrounds etc.) must be kept clean, if necessary by the provision of air curtains etc.

(h) If a wet product is to be sorted and the machine will be subject to periodic cleaning by water, then the optical box must be sealed and some form of optical window provided. Consideration has to be given to keeping this window clean (periodic automatic wash etc.)

(i) The entrance port is usually a simple aperture just large enough to allow for the range of sizes being sorted and the trajectory scatter, but the exit ports may have to be designed to allow for the presence of the ejector and for the scatter of particle trajectories it produces.

(j) Whenever possible, delayed action should be used so that the ejector can be mounted outside the inspection chamber, allowing for better bottom lighting and also minimising the blow back of dirt when the ejector fires. (Delayed ejection is only possible if all particles can be relied on to have the same velocity i.e. the same time interval between inspection and ejection).

The following basic parameters will govern the design of optical chambers:

(a) Size range of particles to be handled.
(b) Number of viewing directions (1, 2, 3, or 4)
(c) Fixed or moveable backgrounds.
(d) Incandescent or Fluorescent lighting.
(e) Internal or external (delayed) ejection.
(f) Wet or dry sorting.

Figure 6 illustrates two typical optical chamber layouts.

8. DESIGN OF FEEDING SYSTEMS

Two basic types of feeding system are used for sorting machines (i) In-line, where particles are presented to the inspection device one at a time, in single file, and (ii) single layer, where a band of particles of appropriate width but only one particle deep is presented. The single layer feed has the advantage of a high throughput but the In-line feed is essential when it is necessary to inspect the whole of the particle surface. Any feeding system has to perform a number of functions:

(a) Metering of the feed -
ensuring that the optimum number of particles per unit time are presented to the inspection system.

(b) Acceleration -
particles require accelerating from rest (in a hopper) to a presentation speed which may be around 250cm/s and which should not vary from particle to particle.

(c) Alignment -
provision of a controlled trajectory through the sensing and separating points.

A great number of systems have been evolved for

meeting the above requirements and a few of the more successful ones are described below and illustrated in figure 7. Virtually all systems utilise a vibrating tray feeder as the metering device on the output of the hopper.

(a) In-line grooved belt -
Acceleration by means of a short chute onto the belt and then by gravity free fall from the end of the belt. Alignment by the belt.

(b) Inclined Chute -
Acceleration and alignment both take place on the chute. Particle velocity at inspection is not so consistent as with the grooved belt.

(c) Inclined contra-rotating rollers -
Acceleration and alignment both take place on the rollers. Reduced friction effects enable a larger range of products to be handled than on the inclined chute, but of course, the capital cost is much higher.

(d) Multi-channel grooved belt -
A multi-channel version of system (a).

(e) Disc-fed belt -
Acceleration by means of rotating disc onto alignment belt.

(f) Flat belt single layer -
Acceleration by means of a short chute onto the belt.

The type of feed system selected will depend on the size range of particles, the type of particles and the accuracy of separation required.

9. DESIGN OF SEPARATING DEVICES

Although a wide variety of separating devices have been tried, those most usually employed are high-speed solenoid valves which release short blasts of compressed air through a nozzle or nozzles. These types of devices have been found to exhibit the essential features of rapid action, reliability and mechanical strength. Other types such as the mechanical gate, involve greater inertia and therefore slower response as well as wear and consequent maintenance problems due to friction and contact with the particles.

The separation process usually takes place while the particles are in free fall and the procedure is to allow the accept particles to continue along their normal trajectory into an accept receptacle whilst the reject particles are deflected out of their normal trajectory by the air blast and are collected in a reject receptacle.

A number of factors have to be considered in the selection of a suitable air valve for use as a separating device:

(a) Choice of materials -
As well as the usual considerations of wear, reliability etc., particular care has to be exercised in relation to the magnetic properties of the material used. A rapid build up and collapse of the magnetic field must be obtainable in conjunction with the associated electronic control circuitry.

(b) Moving Parts -
For rapid response, the inertia of moving parts and their actual movement must be minimised.

(c) Air-flow paths -
The design of the air valve and its associated nozzle must be integrated so to give a

smooth air flow pattern from the nozzle.

The performance of ejector varies considerably depending on the size range of particles that the particular ejector is designed to handle but the following are some typical performance figures:

Particle Size Range (mm)	Operating Pressure (bar)	Air Consumption m³/s	Maximum Operating Speed (Pulses/sec.)
0.5 - 10	2.5	0.05	500
5 - 20	5	0.12	200
15 - 20	5	0.38	50
50 - 150	5	1.50	20

The fastest of the above ejectors is in the form of a solenoid operated disc valve in which the disc movement is as little as 0.1mm. In contrast, one of the larger ejectors consists of two pilot operated spool valves operating in parallel with the high-speed valve forming the pilot stage.

Ejector nozzles vary from a slit 15mm x 0.5mm with the fastest valve to ten 15mm diameter pipes arranged in two rows of five with the slowest valve.

10. SIGNAL PROCESSING EQUIPMENT

10.1. Photomultiplier Circuits

Because photomultiplier tubes are subject to changes in sensitivity due to temperature and power supply variations and also to cater for deterioration in signal due to ageing of lamps, build up of dirt etc., it is usual to employ some form of automatic gain control circuit (A.G.C.) This may consist of a circuit which detects any change in the mean anode current from the photomultiplier (i.e. the background signal) and adjusts the power supply to the photomultiplier in such a way as to restore the original anode current. This circuitry is such that it does not respond to the short duration pulses that result from particles passing in front of the background. In the special case of the aperture disc scanning system (section 6.2) the A.G.C. system works rather differently and in fact adjusts the photomultiplier power supply to maintain a constant signal difference between the background and the chopper disc (black).

The signal corresponding to the passage of particles is taken from the anode of the photomultiplier and if necessary amplified before being fed to the decision making circuitry. In its simplest form, this takes the form of a level discriminator such that all signals of a certain polarity and exceeding a certain noise threshold are classed as reject signals and the corresponding particles as reject particles.

If the scanning type of system is being used, then two discriminators are necessary, the first signalling the presence of the general particle colour and the second the presence of the blemish colour. In this case the two discriminators outputs are separately integrated to give signals proportioned to particle and blemish areas and these two signals are then processed to give a reject signal if the ratio of blemish area to particle area exceeds a pre-set amount.

If a bichromatic separation is being carried out then the two signals, from the two photomultipliers corresponding to the two wavelength ranges, are processed to give a reject signal if their ratio exceeds a pre-set amount.

When a number of detectors are used to examine a particle from a number of directions, then it is usually necessary to provide circuitry to take the decision signals from the various detectors and apply some logic function to them and arrive at a single final decision signal. For example a logic OR function may be used when a reject signal from any of the detectors will result in a final reject decision.

The decision as to whether a particle is a reject or not is reached as the particle passes through the inspection zone but it will not be required to be acted on until the particle reaches the separation zone. This requirement is met by the provision of a delay unit. The first delay units consisted of a magnetic tape loop or drum, but the modern delay unit consists of a solid state shift register. Such units are not only capable of delaying the reject instruction for the required time but also passing information on the size (length) of the particle which can then be utilised in determining the length of time for which the ejector is operated.

The final link in the chain is the ejector drive circuit which provides the necessary coupling between the delay unit output and the solenoid of the ejector valve.

10.2. Solid State Circuits

If solid state detectors are used then it is not usually necessary to provide an A.G.C. system as these devices are inherently more stable than photomultipliers. It is necessary, however, to provide more amplification to bring the signal level up to that of the photomultiplier. This is attained by a low noise, low level current amplifier placed close to the detector so as to avoid interference problems. The decision making circuits, delay unit etc., are as described in 10.1. above.

11. OTHER OPTICAL TECHNIQUES

Apart from the measurement of reflectivity and colour already described, a variety of other optical techniques have been utilised in automatic sorting equipment. A selection of these methods is introduced below.

11.1. X-Ray Fluorescence

This is the most commercially interesting of these additional methods as it has found wide application in the diamond recovery industry.

Basically, it has been found that when a diamond is irradiated with X-rays, it fluoresces, emitting a visible radiation. As the natural gravel in which the diamonds occur does not exhibit this fluorescence phenomenon it is possible to detect the presence of diamonds in a gravel and to separate them out.

A suitable detection system consists of a source of X-rays and a detector which is sensitive to the visible fluorescence but not to the X-rays. If the whole detector system is operated in darkness, the occurrence of a light signal at the detector indicates the presence of a diamond in the viewing position. In practice, because the incidence of diamonds in the gravel is so low and as such large tonnages of gravel have to be processed, the initial sorting is carried out on a broad belt machine and each diamond detected results in the ejection of the diamond together with a number of adjacent gravel pieces. The ejected fraction from the machine can then be hand sorted or machine sorted on an in-line type of X-ray fluorescent sorter.

Utilisation of these types of machine has resulted in a noticeable increase in the diamond production of the mines as a result of the increased efficiency of diamond recovery and possibly due to the improved security of the process.

11.2. Ultra-violet Fluorescence

This method differs from that in 11.1. above in that the fluorescence is excited by ultra-violet radiation rather than by X-rays. Ultra-violet is normally classed as short-wave (around 250nm) or longwave (around 350nm) and different materials behave differently when subjected to these two types of radiation. In practice the shortwave radiation is rather more difficult to handle and has not been extensively used, although a sorting machine for Scheelite ore has been developed.

Longwave ultra-violet on the other hand is proving very useful in the detection of certain moulds and contaminations which occur on food particles and which fluoresces under this radiation.

11.3. Translucency

When a spot of light is focussed onto the surface of an opaque particle a clear image is produced. If a translucent particle is substituted, the image of the spot is surrounded by a diffuse halo of light. This halo is the result of light being transmitted through the interior of the body, suffering multiple internal reflections and then emerging from the surface. If the two types of particle are nearly the same colour, the normal reflection measurements would not be able to distinguish between them but special measurement techniques based on the focussed spot have been developed for this purpose.

Applications of this method include the separation of different types of maize and the detection of quartz-bearing ores.

11.4. Transparency

The separation of transparent particles from opaque is becoming increasingly important, particularly in connection with the recovery of glass particles from municipal refuse. The present method involves the detection of opaque particles by their interuption of a light beam, but problems arise due to optical edge effects, etc., with the irregular shaped glass particles.

11.5. Infra-red

Many optical sorters do in fact have a measurement range which extends into the infra-red as both the light source and detector may cover both the visible and near infra-red ranges of the electro-magnetic spectrum. It is therefore perhaps not inappropriate to mention one pure infra-red method which is now finding application in automatic sorting. This method involves the remote sensing of surface temperature utilising infra-red detection techniques. It is anticipated that this will have important applications in the mineral field where the surface temperature of a pre-heated ore is indicative of its constituency.

12. SIZE AND SHAPE MEASUREMENT

The measurement of size and shape of regular shaped objects by electro-optical methods is a well developed technique. However, the applications of these measurements to irregular shaped particles such as are found in the food and mineral industries presents a number of problems, not least of which is the definition of what is meant by "size" and "shape".

An excellent example of the problem of shape definition and measurement is afforded by the tomato industry in America where shape is one of the factors which determines the grading and therefore the value of a tomato. The authorities issue regulations and models to try and define shape classification but even so, the final result seems to depend to a great extent on the interpretation of the individual inspector, and so far it has not been possible to classify shape on the basis of readily obtainable measurements.

Equipment has been developed which utilises arrays of photodiodes to give measurements of the height and several of the cross-sectional diameters of the fruit but a method of correlating these measurements with shape grade has still to be found.

In many sorting applications we are concerned with the measurement of a surface concentration, that is we wish to know what percentage of the surface has a particular colour. This measurement has been achieved in some machines by use of the mechanical scanning camera and in others by the scanned photodiode array (section 6.2) One application where neither of these methods could be applied was the ultra-violet fluorescence detection of Scheelite ore (section 11.2) In this case it was necessary to determine what percentage of the total surface area was exhibiting fluorescence, but as the operation had to be carried out in darkness the problem was, how to measure the area which was not emitting fluorescence. This problem was solved by utilising a linear array of infra-red emitting diodes placed at one side of the optical chamber and focussed onto a single infra-red detector at the opposite side. The infra-red emitting diodes were pulsed on sequentially and then as the particles fell between the diodes and the detector it was only necessary to count the number of pulses that did not reach the detector to determine the cross-sectional area of the particle. The use of infra-red ensured that this radiation did not interfere with either the ultra-violet radiation or the visible fluorescence.

13. CONCLUSION

It is hoped that this paper will have given mechanical engineers at least a broad understanding of the principles involved in the incresingly important field of automatic colour sorting.

Many challenging mechanical problems still remain to be solved and involvement in future projects will no doubt prove very rewarding.

14. ACKNOWLEDGEMENTS

The author wished to thank the Directors of Gunson's Sortex Limited, for permission to publish this paper and to acknowledge the help given by Mr. T. H. Chapman in its preparation.

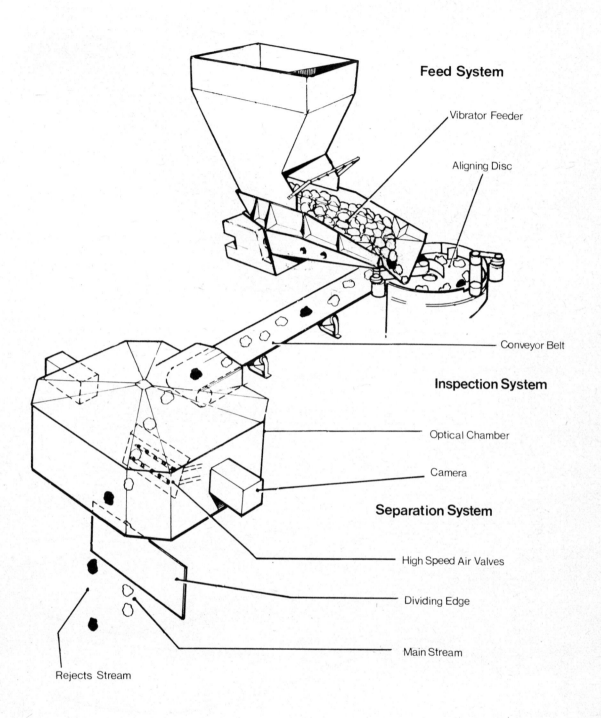

Feed System

Vibrator Feeder

Aligning Disc

Conveyor Belt

Inspection System

Optical Chamber

Camera

Separation System

High Speed Air Valves

Dividing Edge

Main Stream

Rejects Stream

Fig1. Typical Optical Sorting Machine

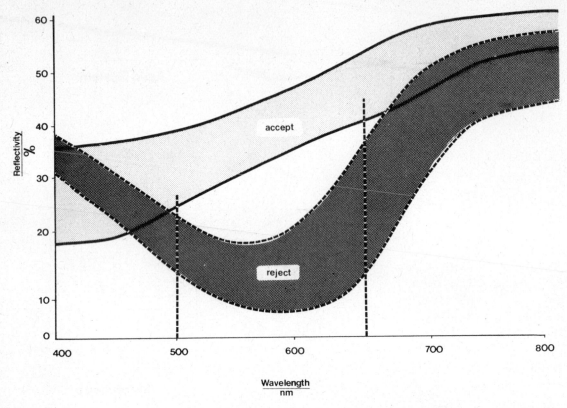

Figure 2   Spectral Curve - Monochromatic

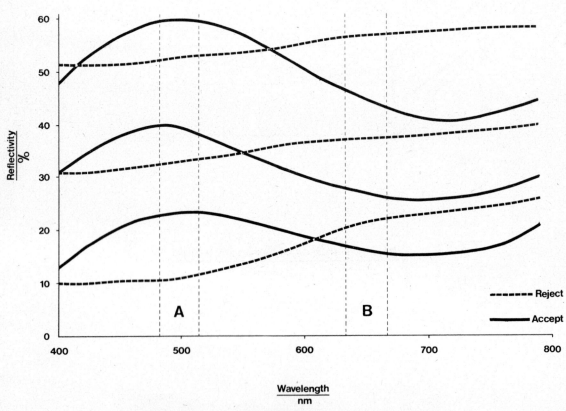

Figure 3   Spectral Curve - Bichromatic

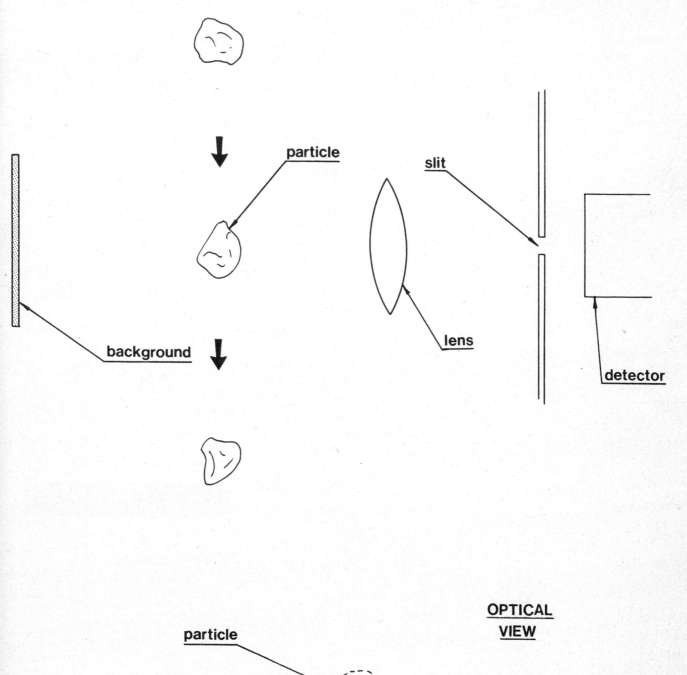

particle

slit

background

lens

detector

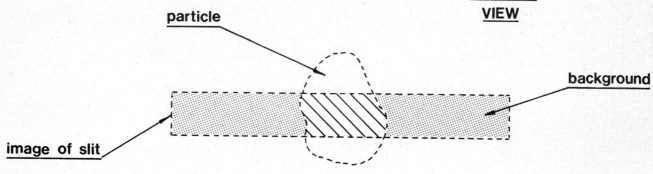

OPTICAL
VIEW

particle

background

image of slit

Figure 4   Matched Background System

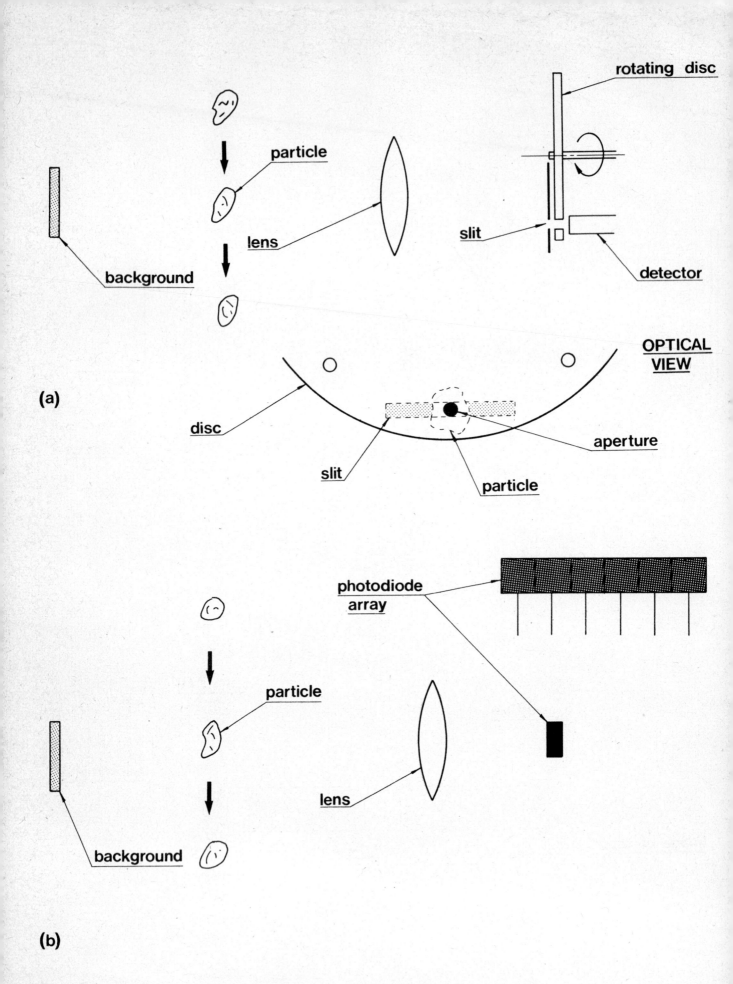

(a)

rotating disc

particle

lens

slit

detector

background

OPTICAL VIEW

disc

slit

particle

aperture

photodiode array

particle

lens

background

(b)

Figure 5   Scanning Systems

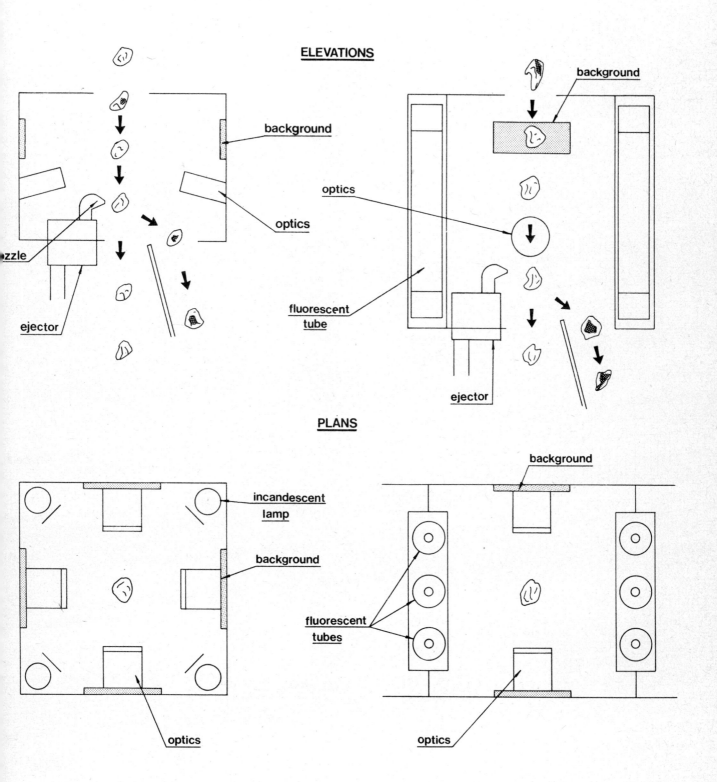

Figure 6   **Optical Chamber Layouts**

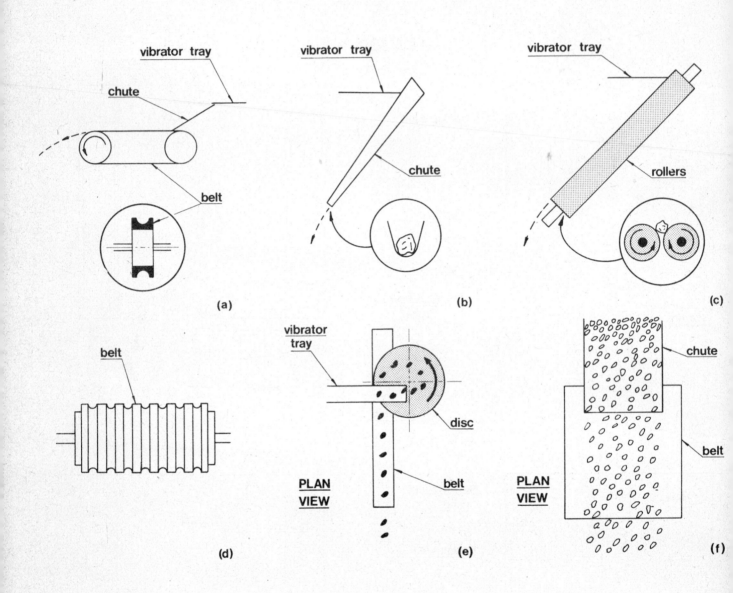

Figure 7   Feeding Systems

C126/74

THE USE OF LUMINESCENT MATERIALS IN MECHANICAL HANDLING SYSTEMS

MICHAEL SILVERTHORNE AGATE, MIEE
Group Head, Postal Headquarters, Mechanisation & Buildings Department,
Letter Division, PMB 143, Leith House,
47 Gresham Street, London EC2V 7JL
The Ms. of this paper was received at the Institution on 31st December 1973 and accepted for publication on 28th February 1974. 33

SYNOPSIS   The requirements leading to the use of luminescents and the choice of phosphors in particular for postal coding are examined, and an account is given of phosphor coding practice as applied in the British Post Office letter handling systems.   The paper concludes with brief reference to recent developments and applications in industry.

INTRODUCTION

1   Luminescent materials are now used all over the world in letter handling systems and most of this development was pioneered in this country. The range of materials which could be used for this purpose has been reviewed in detail, reference 1, but some comment on the factors which led to the development of postal phosphors is appropriate.

THE CHOICE OF LUMINESCENT MATERIALS

2   The postal envelope gathers wanted and unwanted marks even under manual processing and more so with mechanised conditions. Some of these marks may be highly prized by collectors and nearly all are inviolate, but for the engineer the unwanted marks form the noise environment in which his system must work; it is essential therefore that the materials adopted for coding must provide a good signal to noise ratio. A direct optical system employing an idealised black mark on clean white paper can only offer a theoretical signal to noise ratio of 24 db. Magnetic materials were considered and we know these have been highly successful for encoding documents in computer controlled systems, notably cheque coding, but the mail system with its extreme range of paper qualities, together with size and thickness variations, lacks the controlled conditions necessary for success with magnetic coding.

Preliminary studies showed that luminescent materials afforded signal to noise ratios in the order of 26-34 db and with a tentative decision to use this technique the choice lay between instantaneous and delayed photo-luminescence. Fluorescent materials are cheap, harmless, and easy to apply and handle, being available in liquid form. They emit light only during irradiation, consequently optical filters are required to distinguish between the wanted signal emission from the fluorescent code marks and background "noise" due to the UV source. These technical problems can be overcome but since most varieties of paper used in stationery themselves fluoresce to a certain degree, detection of fluorescent markings resolves into discrimination between degrees of fluorescence. Today, the ubiquitous optical brighteners in envelope papers have seriously reduced the signal to noise ratio and embarrassed those administrations that still use fluorescent materials for address coding.

The choice led inevitably to materials exhibiting delayed photo-luminescence, known collectively as phosphors. These materials fluoresce but continue to emit light after removal of the source of irradiation for periods which can vary from a few milli-seconds to seconds; during this afterglow period conditions are excellent for reading signals against the darkness of a "zero noise" background.

CODE-SORTING REQUIREMENTS

Although phosphors are more reliable from the standpoint of signal to noise ratio the inorganic phosphors at first available were expensive and could not be ground to a pigment size suitable for application by printing methods without considerable loss of optical efficiency.

Attention was, therefore, turned to overcoming the deficiencies of the existing phosphors and producing a phosphor material which was cheap, non toxic, easily applicable to paper and waterproof. Post Office chemists developed a number of organic phosphors in resin form. These could be ground to a fine powder, suitable for incorporation in media to enable transfer to paper by printing methods, reference 2.

A complete family of organic phosphors was eventually developed followed by the development of a suitable code printer. There is a marked tendency to sedementation in inks formed of solids in suspension in liquid media. The higher the mass of the particles the worse this problem becomes. A typical phosphor has a particle size range of 2-25 microns with an average size of 11 microns which severely restricts the range of printing methods possible. Fortunately there was one printing process which overcame this problem, namely the old method used by bookbinders for applying gold leaf to book covers, known generally as blocking transfer printing. An account of the objectives of the code-sort system, the development of the first coding desk and associated printer is given in reference 3.

Further research led to the development of new phosphors whose optical performance was characterised by a very low response to irradiation at 365 nm, but responsive in the normal way to the shorter wave length 254 nm irradiation, such a phosphor can thus be used to provide a second and independent coding method which can exist and function side by side with the existing code sorting phosphor provided the short wave response phosphor is used before the long wave one is applied (the latter responds to both irradiation wave lengths).

Table 1 gives the code letters for the principal phosphors which have been developed and used by the Post Office together with the activator and the optical characteristics in each case.

POSTAGE STAMP CODING

3   All automatic facing machines are based on the principle that the stamp is invariably placed in the top right hand corner of the front of the envelope and can therefore be used to identify the position of the address.

The stamp itself with its characteristic colour for each value, can be used as a trace to determine both letter orientation and code without additional marks. A Japanese facing machine uses this principle successfully by breaking up the light reflected from the stamp into the three primary colours in order to assess the stamp value by vector summation of the outputs from three photomultiplier detectors. However, the British Post Office considered that this system has too great a restrictive effect on stamp design, especially when associated with tariff changes, and rejected it in favour of the unique trace element provided by the use of phosphor coding.

The ALF machine

The process of orientation for address reading is known as facing, and the equipment designed to do this work is called the Automatic Letter Facing machine (ALF), and has been described in detail, reference 4. In this machine the facing of letters and sorting them into two categories of mail priority, 1st and 2nd class, is directly controlled by the phosphor stamp code.

The trace element

Facing is achieved by scanning every letter for the presence of the phosphor coded stamp in each of four possible stamp corners. The machine is designed to operate at a fixed speed of 100 m per min and a single belt carries the letters closely spaced through it in one continuous process. In the irradiation section special belts allow almost all of both sides of the letter to be irradiated. The ultra violet lamp is of the TUV (germicidal) type which is a low pressure mercury vapour lamp; the principal emission is at a wave length of 254 nm. Lamps are positioned on either side of the letter path and their length, 400 mm, and the speed of the letter, determine the irradiation time, 240 mS.

The phosphor scanner uses the highly sensitive photomultiplier as a detector, and the letter, travelling past the scanner block has its face pressed closely against the scanner window to exclude all stray light. Detectors are mounted immediately following the irradiation section, as close to it as belt configurations will allow. However, inevitably all of the four scanners cannot be sited equally advantageously with the result that successive scanners must be increasingly sensitive, as their distance from the UV irradiation increases. These conditions are illustrated in Fig 1 which shows the relative light emission in relation to time as the letter passes through the detection area; the exponential nature of the growth and decay is evident.

The physical presence of a letter travelling through the scanner area is also monitored by photo-electric detectors and the information about each stamp scan is entered into a shift register, together with the letter position information. Interpretation logic circuits control the appropriate divertors to carry out the reorientation functions, reference 5.

The code element

Separation into tariff classes is carried out after facing has been completed and the scanning units to achieve this are very similar to those already described. The code arrangements consist of over-printing the stamp with TPA phosphor, illustrated in Figure 2. The code for the minimum 2nd class tariff consists of a single vertical bar at the centre of the stamp; only one stamp is so coded. All other definitive low value stamps carry two vertical bars in the positions shown, first class codes are thus applied to the stamps which are for 1st class tariff or more, and to lower values (known as make-up values) which could be used to increase the second class stamp value to 1st class rate. Certain variations of these arrangements are liable to occur with special issue stamps when artistic requirements dictate.

Separation is by assessment of the duration of the stamp signal. The 2nd class stamp code element having a width of 4 mm and travelling at 2.36 mm per mS is scanned by a slot 4.4 mm wide which generates a pulse not exceeding 5.4 mS duration and rather less than this at the logic threshold level. Any signal less than 7 mS duration is interpreted 2nd class. 1st class tariff codes and combinations of codes always generate a signal which persists for more than 7mS while the stamp corner gates are open. The decision for each letter is entered into a shift register and accompanies the letter electrically until it arrives at the appropriate point for diversion to the stacks.

Printing phosphor stamp codes

Most stamps for Great Britain are printed by the photogravure process. A copper cylinder is photo-etched with a cell structure which, when charged with ink, transfers its impression by a very high pressure applied between the roller and print paper. The process is capable of a high degree of definition and this in turn requires a high surface finish paper; there is usually no difficulty in controlling the amount of phosphor ink printed, or its registration with the design. A TPA phosphor code is printed on every stamp issued and continual development has brought this to a low cost.

Quality Control

Currently the production requirements of postage stamps with phosphor codes is 7000 million stamps per year. Sampling procedures for the inspection of the incoming basic phosphor from the manufacturers, the ink media and diluents, and the made-up batches of phosphor ink are all carefully carried out and recorded against stamp sheet numbers.

Daily samples of printed stamps are measured using the Phosphoresence Tester reference 6 which enables quantitative assessments to be made of individual stamp samples under any required conditions of irradiation and reading delay. This measurement procedure, whilst providing a useful laboratory check programme, clearly cannot give effective quality control of such large scale production as occurs in stamp printing, and as a result, in spite of the very high quality of stamp printing, overprinting of the phosphor has a somewhat irregular quality record. Many factors affect the phosphor print quality and are difficult to control when there is virtually no visual feedback.

On-line Phosphor Printing Monitor

At the present time monitoring equipment is in the final stages of design and development, which will provide on-line supervision of the phosphor code mark printing on the stamp printing machine.

The printed web is irradiated with uv light at three places across its width, the uv sources being provided with detectors to monitor and control their intensity. A photomultiplier detector unit is located "downstream" and adjacent to each irradiation position to monitor the level of luminescence from the printed stamp code strips. The speed of the printing press is also monitored at a suitable point, and the analogue signal so obtained used to generate a simulated decay curve representing the phosphor decay characteristic at the operating speed of the printing machine; schematic diagram figure 3 refers.

The two signals from the outer phosphor monitor heads, situated at the left hand and right hand of the print web are compared directly to provide a means of achieving print balance across the web; the centre signal is continuously compared with the derived speed-dependant reference standard. Digital indicators display the light emission level and visible and audible warnings are provided at predetermined warning and action limits.

Performance of the stamp in the sorting office

The Post Office carried out a study to assess the effect on machine performance of stamp colour, tone, and the amount of phosphor applied, reference 7. A specially selected series of stamps of four colours was overprinted with phosphor density (grams/m^2) in six steps descending in geometric progression. When made up as test letters they were used to establish the fail point of the machine functions in relation to these variables.

The Postage stamp requires a high degree of engineering control if it is to be operationally successful. To reconcile the sometimes conflicting needs of artistic expression with operational requirements is often a difficult and delicate matter. Because of its monochromatic nature the light emission of a phosphor signal is dependent on the colour of the stamp on which it is printed. In the case of very dark and spectrally unfavourable stamp colours, reference 7, absorption is nearly total and the signal which remains to control the machine comes mainly from the white perforated surround of the stamp, and thus only about one tenth of the phosphor printed area is effective. Continual study by Post Office chemists has enabled empirical standards to be set to explain this phenomena, and whilst in no way restricting stamp design, provides a useful guide to explain engineering requirements to stamp designers.

POSTAL ADDRESS CODING

4   Address coding is essentially an information encoding system and forms a fundamental part of the principle that each letter address should be read once only within the sorting process. The postman at the coding desk copies the postcode into the system using a typewriter keyboard; without the postcode he must extract the information from the address. The address information, now in 5 unit signal code, is referred to a translator and the appropriate 24 bit code is sent back to the coding desk of origin to be stored until the printer cycle initiates the "pin setting" process, after which printing takes place. The letter thus coded, may now be "read" and sorted automatically in successive stages by any sorting machine in the country.

Printing

The printing method used in the British Post Office address coding system is adapted from a process which has been in existence for many years. A heated pin is applied to the letter with a special blocking (transfer) foil carrying the phosphor interposed, and a light pressure applied. The conducted heat melts the carrier compound which adheres to the envelope surface and freeze dries almost instantaneously.

Paper

By this is meant envelope paper and letters in general. It is certain that no traditional press type of printing process could deal with the extremes of quality range encountered in the post, in terms of surface finish and porosity. In addition the thickness of letter items at present regarded as machinable extends from 0.25 mm, the specified minimum for cards, to 6.5 mm as a somewhat arbitrary maximum for letters; at the upper limit, items can be inflated with air especially in the early stages of machine handling. Blocking transfer printing is able to deal with these characteristics with something to spare and give a consistent performance.

Drying time

Most printing presses are designed to allow a considerable drying time before the machine interferes and would smudge the print. The mail handling machine, because of the weight range of letter items and the effects of machine generated static, is in general designed to hold the letter at all times and a long section for drying print would be uneconomic. The process chosen does not in fact involve "drying" as such by evaporation but consists of the virtually instantaneous setting of a hot melt wax on the cold surface of the envelope at the moment the heated pin is withdrawn. The setting time is believed to be less than 500 ms.

Blocking foil manufacture

The requirements of the phosphor transfer foil, or luminescent tape to give it its Post Office title are complex. The problems involve achieving an accurate and homogenous coating with the "heavy" phosphor, due to its particulate nature, and so coating the substrate that the working mix fully parts from it and attaches itself to the letter material. When this occurs correctly a precisely controlled amount of phosphor is deposited on the letter and this contributes considerably to simplifying quality control of print in the sorting office. The relegation of this responsibility to the tape coating manufacturer makes good sense from the operational point of view but the high cost of tape forms a significant part of the cost of the printing process.

Print pin temperature

A typical print temperature characteristic is illustrated in Fig 4, each point being the mean light emission from 350 tests at the temperatures shown, with pin dwell time of 100 ms and a pin force of 6.6 N. This shows that effective transfer under these conditions takes place at temperatures above 110°C.

Of necessity, the two parameters pressure and dwell time are built into and fixed by the printer design, and for some years have been standardised at 4.4 N and 140 ms. Temperature is standardised at 130°C which requires an oven temperature of 155°C, the latter being controlled by a thermostat to $\pm$ 5°C.

Code Mark Format

The code mark format is becoming a familiar part of the postal letter item and has been illustrated, reference 8. Each part of the two part code consists of a 12 bit binary arrangement with a start mark always present, and at the opposite end a parity mark, to make the total marks in each row an even number. The outward sorting code is printed along the bottom of the face of the envelope 10 mm from the long reference edge, and a similar code arrangement for inward sorting, is situated 60 mm from the same reference, which for most circumstances allows the address to appear in the space between. The code elements are 3.17 mm diameter at a pitch of 6.35 mm giving a 1:1 mark : space ratio. This rather coarse structure is necessary because of the print and paper variables. It would be an advantage if techniques now becoming available could be used to print a much finer and more closely spaced format, say five bits per cm. This would enable the full 24 bit machine code to be printed along the reference edge of the shortest letter envelope and avoid some problems associated with window envelopes, stamps and other marks.

The sorting machine reader

The reading of address codes takes place in the sorting machine. Irradiation is by a TLD type lamp, emitting uV at 365 nm. The length of the lamp is 230 mm and the nominal letter transport speed is 58 m/min. A single photomultiplier scanner unit is situated 155 mm from the end of the irradiation section and these dispositions result in an irradiate time of 250 mS and a read delay of 170 mS.

Timing signals derived from the letter transport, control the operation of a strobe and shift pulse generator, the purpose of which is to strobe or interrogate each code mark position eight times at 0.4 mm intervals and produce a shift pulse which moves the code mark information along the reader shift register, between each code mark position. The entry-exit photo-electric detector is situated 6.3 mm ahead of the phosphor scanner and as the letter cuts this, the logic circuits are primed to receive the start mark information, which is printed a nominal 6.3 mm from the leading edge of the letter. When the start signal is received the reader logic processes are iniated

and all strobe pulses are timed from this transition including the 14 stage reader shift register; the logic requires two out of eight strobes in order to enter a valid mark decision into the store and this provides a high working margin for the many variables which can affect the letter and its printed code.

Quality Control

Much effort has been spent in 'designing out' the variables in the coding and reading processes, and even the number of adjustments available operationally. A standardised manual printer has been designed for use with the luminescence tester and this enables QA assessments on luminescent tape to be carried out at manufacturers' works.

By far the most effective control of the process within the office is afforded by Letter-Office Computerised Monitor equipment, LOCUM, reference 9. A computer is used to build up a quantitative picture of the code printing and reading system, with individual assessments of the strobe counts of address code mark prints under limiting conditions, such as print temperature variation, print on window envelope materials, print over stamps and cancellation marks, and detailed variations in printer design as mark numbers are advanced. The data obtained from LOCUM studies has been invaluable for machine designers, maintenance engineers and operational managers.

NEW DEVELOPMENTS

5   "All-over" phosphor stamp coding

A number of experimental issues of stamps have been made in which the luminescent material has been incorporated in the chalk and clay surface coating normally applied to the paper to provide a high grade printing surface. The method should result in a cheaper product but will also bring significant advantages in process and quality control which will reduce the risks of occasional complete failures. The process will be applied to all stamps except the minimum 2nd class tariff which will continue to be printed on untreated paper; the reader logic remains unaffected.

Electro-static jet printer.

This new form of printing (or writing) was developed for oscillography by R G Sweet in 1963, reference 10.

A stream of liquid ink is formed into a fine jet and broken up into precisely controlled droplets by a transducer driven at 100 kHz. As the droplets leave the jet they pass through a ring from which each ink particle receives a charge in accordance with the amount of deflection required. The charged particles are subjected to a fixed electro-static field which deflects them in a controlled pattern of code marks or characters, formed as they strike the paper.

Equipment using this principle has been developed and is in use commercially in this country for banking document coding and accounting. In this application a red fluorescent ink is used which requires technical control of the document paper quality, especially luminescent content. Phosphor ink, because of its hitherto particulate characteristic has not been suitable for jet printing, and although the system has been used for postal purposes in America, using a simple optical system, the British Post Office has continued with the development of a non-particulate phosphor for jet printing applications.

Ink jet printing offers considerable advantages because of the complete absence of moving parts, and no physical contact with the object to be printed, enabling high speeds to be achieved. The normal mode for jet printing postal address codes would be to print "on the run", executing the code elements sequentially instead of simultaneously; it is expected that letter throughputs of 10 items/sec at transport speeds of 3m/sec could readily be achieved. It is not yet known to what extent this technology, in conjunction with the liquid phosphor under development, will succeed in dealing with the wide range of paper quality and other variables in the letter post.

Non-particulate phosphors.

Post Office chemists, continuing the study of organic phosphors have now reached an advanced stage of development of a process for making a new phosphor in liquid form. When used in some of the current designs of jet printer, it matures on the paper substrate a few minutes after printing, and can be arranged to exhibit the characteristics of either of the presently used TPA and CSA phosphors.

INDUSTRIAL APPLICATIONS OF CODING TECHNIQUES

6   Coding of mass produced items has been practised for many years and is a vital part of communication in larger organisations. The Post Office Vocabulary of Engineering Stores for example has over half a million items, each having a code number for computer control of purchasing, stock and issuing operations; in addition, every pack, and sometimes every item, is marked with an alpha-numeric code, identifying the manufacturer and year of manufacture.

Not all coding is quite so overt or straightforward in purpose, and many manufacturers items are coded in a way that is meaningless to the public at large; this applies in particular to processed foods where codes are used for the confidential control of deterioration. Luminescents are already used in a trial scheme for price coding goods in supermarkets, providing direct read facilities at the cash desk.

The packaging industry would appear to offer the most likely field for coding techniques using luminescent materials. The Post Office has been consulted on many projects, particularly in applications where it is important that goods shall be accurately associated with their correct package and description, as in the case of drugs and medicines. Here it was thought that unobtrusive coding could be used to provide a continuous monitor on packing and labelling operations. However we find that in general the problems of applying code marks and reading them with irregular shaped objects, many of them in bottle or cannister form, which rotate in a random fashion in the machine have proved too difficult using luminescents, it is to be hoped however, that the new jet printing technique, together with new phosphor materials may sufficiently simplify the application problems and make luminescents practicable

for wider industrial applications.

Many adhesives used in packaging are inherently luminescent or could be made so with additives, thus light detection after irradiation forms a useful method of inspecting the glue line on packaging materials before assembly into cartons. This technique may well be justified in cases where failure of the glue-line before or during the assembly process could cause serious jams in the machine followed by costly down-time.

Luminescents are used with ordinary printing processes to provide a security check on documents. One example is a well known trading stamp concern which uses phosphor coded stamps for an independent gift scheme. Security coding in this way can be extended by using two phosphor activators, one of which requires a particular irradiation wavelength to check the emission. Further, detection equipment capable of determining emission wavelength can be used in conjunction with selective phosphors. All such schemes provide a measure of semi-annonymous security, depending for their effectiveness on the difficulty of procuring the necessary technology.

CONCLUSION

7   Efficient coding systems have been developed by the Post Office, and production versions in the field have proved themselves to be effective and reliable in controlling letter handling machines with adequate working margins for the exceptional variability which prevails in the product handled. New developments in printing techniques have been mentioned, the scope of which has not yet been explored, but which it is suggested could lead to the wider use of luminescents in the packaging and handling industry.

ACKNOWLEDGEMENT

8   The author wishes to acknowledge the permission of the Director of Mechanisation and Buildings, Post Office, for permission to publish this paper.

APPENDIX 1

References

1   Forster C F. The use of phosphorescent code marks in automatic letter-facing and sorting machines. PO Elec. Engrs. Journal Vol 54 1961.

2   British Patent No. 870504.

3   Pilling T and Gerard P S. Automatic letter sorting - The Luton Experiment. PO Elec. Engrs. J. Vol. 54 pp 31-36 1962.

4   Copping G P, Gerard P S and Andrews J D. Automatic facing and stamp cancelling. PO Elec. Engrs. J. Vol. 53 p12 1960.

5 · Wicken, C S. Control systems for ALF. Proc. I. Mech E. Vol 184 Pt 3H. Paper No. 18, pp 95-100 1969-70.

6   Harrison J C, Rickard E F, Forster C F and Walker A D. Proc. I Mech. E Vol. 184 Pt 3H. Paper No. 4. pp 52-59 1969-70.

7   Agate M S. Assessment of the performance of phosphor postal items. Proc. I. Mech E. Vol 184 Pt 3H pp 60-66 1969-70.

8   Andrews J D. Coding principles and practice.
Proc. I Mech E. Vol 184 Pt 3H. Paper No. 12 pp
139-145 1969-70.

9   Andrews J D, Bennett H A J and Pratt A D.
Letter office computerised monitor - LOCUM PO
Elec. Engr. Journal Vol. 65 pp 172-176 1072.

10   Sweet R G. High frequency recording with
electrostatically deflected ink jets. Review of
Scientific Inst. Vol 36 No. 2 pp 131-136
Feb 1965.

LIST OF ILLUSTRATIONS

Table 1

Dates used	Use	Code	Activator	Irradiation Wave length nm	Light emission colour	Half Life mS (1)
1959 to 1961	Stamps	PHD	para-hydroxy diphenyl	254	Blue-green	500
1961 to 1965	Stamps	CSA	carbozole sulphonic acid	254 and 365	Blue	480
1965 on-	Address codes					
1965 on-	Stamps	TPA	tera-phthalic acid	254	Violet	125

Note 1   Half life calculated from the earliest measurable value, ie 40 mS after irradiation.

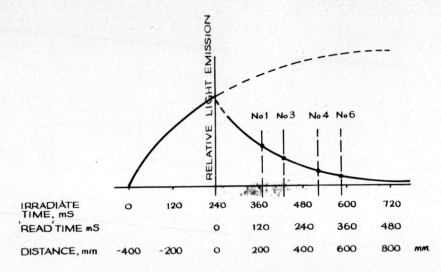

FIG.1 ALF SCANNERS - SHOWING
RELATIVE LIGHT EMISSION/DISTANCE
RELATIONSHIPS (TPA PHOSPHOR)

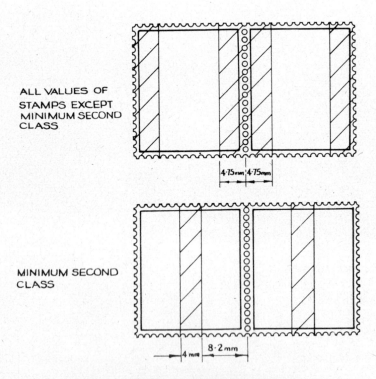

ALL VALUES OF
STAMPS EXCEPT
MINIMUM SECOND
CLASS

MINIMUM SECOND
CLASS

FIG.2   POSITION OF PHOSPHOR OVERPRINT
BRITISH LOW VALUE STAMPS

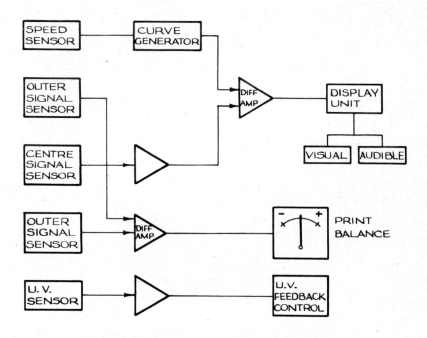

FIG.3. SCHEMATIC DIAGRAM OF ON-LINE MONITOR
EQUIPMENT FOR PHOSPHOR STAMP CODE
PRINTING.

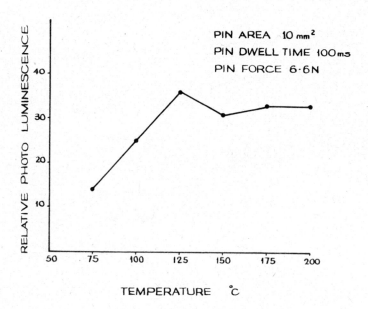

PIN AREA 10 mm^2
PIN DWELL TIME 100 ms
PIN FORCE 6·6N

FIG. 4   CODE MARK PRINTING SIGNAL/
TEMPERATURE CHARACTERISTIC

ELECTRO-OPTICAL EQUIPMENT IN PAPER-MAKING AND CONVERTING

ROBERT JAMES McGILL, B Sc F Inst P, M Inst MC
Systems Engineer, Wiggins Teape Limited,
Stoneywood Works, Bucksburn, Aberdeen AB2 9AB
The Ms. of this paper was received at the Institution on 10th December 1973 and accepted for publication on 28th February 1974. 33

SYNOPSIS.   Papermaking has made good progress in adapting to modern technology.   One feature of this has been the use of electro-optical devices in the mechanical handling stages of the process.   Minor uses include web guiding, counting and watermark registration control.   The most spectacular progress, however, has been in the application of such equipment for detecting defects: together with the development of suitable mechanical handling machinery, this has enabled the sorting process - the removal of defective sheets - to be made automatic.   Many successful machines are in operation.   Increase in the number of such applications depends on economic and operational factors.   Further progress may lie in the extension of the use of inspection equipment throughout the process as an integrated system.

INTRODUCTION

1.   Papermaking is an ancient craft which developed in step with contemporary technology.   The industrial revolution saw its conversion to a continuous process. Fundamentally, the process is substantially the same today.   The technological revolution of the last quarter century has produced dramatic changes in method and performance. Scientific techniques have replaced many of the human skills.

Electro-optical equipment has made an important contribution to the upgrading of efficiency. Applications in the mechanical handling stages of the process include web guiding, counting, registration control and the detection and handling of defects. Of these, the last mentioned is by far the most prominent, complex and costly and has produced the most revolutionary changes: for these reasons, this application forms the main subject of this paper.

PAPERMAKING

2.   Paper is a heterogeneous material, composed principally of cellulose fibres. It is produced on large machines as a continuous web.

The general description "paper and board" covers a very wide variety of products.   The details of the process differ accordingly. Figure 1 is a simplified block diagram of a basic system.

The mechanical handling of paper as such begins after the papermaking machine. The final stages come under the general heading of "finishing":   it is in this area that most of the applications of electro-optical equipment, relevant to this conference, are to be found.

MISCELLANEOUS APPLICATIONS

3

3.1 Edge guides

In certain finishing operations, such as slitting, the web must be accurately aligned with the machine in which it is running.   Electro-optical devices are commonly used to detect and control the position of the edge of the paper. The sensors used are mainly photoelectric units, varying in complexity according to accuracy requirements.   Control is effected by moving the whole unwind unit transversely or changing the angular position of one or more carrying rolls.

3.2 Counting

Automatic counting is often performed on the cutting machine by photoelectric detection of the individual sheets as they are conveyed through the machine.   Electronic or electromagnetic batch counters are used and the batches identified by the automatic insertion of tabs.

3.3 Register Controls

Some watermarked papers are required to be cut with a fixed relationship between the positions of the watermark and the cut line.   The required position of the cut line is identified by impressing a mark on the paper on the making machine, as part of the watermarking operation.   The actual cut line must pass through or near to this mark. This is controlled automatically by sensing the position of the mark by photoelectric means, actuating mechanisms which adjust the relative speeds of the paper and the rotary cutting knife.

DEFECT DETECTION AND SORTING

4

4.1 General requirements

For present purposes, defects are defined as

localised irregularities of an undesirable kind.

Defects may originate in the raw materials or may be generated within the process. Precautions are taken to minimise them but it is impossible to eliminate them altogether. Normally, they represent a tiny proportion of the total product, by area or volume, usually less than one part per million, but their presence can seriously affect relatively large amounts of otherwise satisfactory material or cause expensive process losses. A single flaw can cause rejection of a sheet of paper which is 99.99% perfect, or, much more serious, it can cause costly shutdowns in the manufacturing process or, still worse, in the user's process, in which case heavy compensation payments may have to be made.

For any product, some form of inspection is desirable, to protect outgoing quality. In some cases, the value or importance of the individual item justifies complex and costly inspection. Paper is at the other end of the scale: the selling price of the product unit, a single sheet of paper, simply cannot justify more than a very small inspection cost. It is only on the highest grades of paper that 100% inspection can be justified.

4.2 Manual sorting

The traditional method for full inspection of sheeted paper is to employ a small army of girls who manually handle and inspect each sheet and separate the good from the bad. This operation is known as sorting. Less expensive methods, such as random sampling and "fanning" through a stack, are also used, but are not wholly satisfactory.

In full manual sorting, only a second or two can be allowed to handle and inspect the sheet on both sides. A snap decision must be made, comparing the visual impression with a mental picture of accept/reject levels, acquired during training. These levels are seldom clear-cut and each decision is based on an overall assessment. Each girl is expected to maintain a standard performance throughout the working day. It is not surprising that quality levels vary. Performance differs between individuals and is affected by alertness, fatigue, etc. Momentary inattention can and does allow gross faults to pass.

The human capability for this kind of work is high in discrimination and versatility but low in uniformity and reliability. Other disadvantages of manual sorting include the high cost of labour, providing amenities, etc., difficulty in obtaining labour, also the cost of damage to the product caused by handling and transport.

The need for automatic methods has long been obvious. The possibilities opened by advances in electronics led to the development of equipment for the detection of defects. At the same time, suitable mechanical handling machinery had to be developed.

4.3 Equipment requirements

In making the transition from manual to automatic inspection, it is necessary first to specify requirements in precise terms. Accept/reject levels must be defined. This is a difficult problem because of the wide variety of product types, qualities and end-uses. A general definition simply is not possible. It is difficult to obtain comparable standards even on similar or identical grades. Agreement is possible only at the extremes, i.e., on the very bad and the very good. Between lies a very wide range of types, sizes, and frequencies of defects, on which decisions are affected by subjective considerations, individual idiosyncracies, the state of the market and so on.

We can make a start by classifying defects as (a) those which are of a mechanical nature, such as tear-outs, folded corners, holes, lumps, joins, creases and similar deformations and (b) purely visual blemishes, such as dirt spots, discolorations and marks of various kinds. Gross examples of the mechanical type are obvious potential sources of process troubles. They are sometimes called "press stoppers". Small mechanical defects may also cause trouble: some of these are very difficult to detect. Visual defects are usually undesirable because they spoil the appearance of a sheet, only becoming serious if they occur in large numbers. However, in some papers, e.g. those used in optical character recognition systems, visual defects may be as serious as the mechanical kind.

For each type of fault, a decision has to be made on accept/reject levels, according to seriousness, size, numbers, etc. This is largely a matter of subjective judgement.

An inspection system must be tailored to suit individual requirements. Allowance must be made for the fact that the machine has no subjective judgement. It is low on discrimination. It has the unpleasant habit of doing what it is designed to do, rather than what one wants it to do!

Apart from its functional requirements, inspection equipment must satisfy practical conditions. This is vital if it is to survive as a viable part of the process.

It must be rugged, to stand up to arduous conditions, including a certain amount of misuse.

It must be simple to operate, with controls limited to essentials, easily understood, without ambiguity. It is operated by people trained to handle heavy process machinery and unlikely to be "instrument minded".

Performance must be stable, requiring no readjustment or trimming, over reasonably long periods, unaffected by its environment, e.g., by paper dust, a common hazard.

It must be reliable, capable of operating continuously, without breakdown, for long periods.

It should be simple and inexpensive to maintain and repair. The level of technical ability required for maintenance should not be higher than is normally available in a paper mill.

Finally, it is preferable that the equipment is compact, to minimise obstruction to normal operation.

4.4 Web coverage

The proportion of the total paper surface inspected differs according to requirements.

Coverage in the running direction ("machine direction") is simply performed by the passage of the moving web through an inspection unit, located in a fixed position on the processing machine. For some purposes, inspection of a narrow strip is sufficient; for others, extended coverage in the transverse direction ("cross direction") is provided by making a single inspection unit suffic- iently wide, or by using a multiplicity of units, or by a scanning arrangement.

Inspection of only one side of the sheet is often adequate. Holes, creases and large in- clusions are common to both sides and the end-use may not justify two-sided inspection for surface dirt only. Also, where optical inspection is made by transmitted light, surface dirt on both sides is detectable. In some cases, however, separate optical inspection of both sides justi- fies the greatly increased cost.

4.5 Inspection equipment   (Figure 2)

4.5.1 A variety of equipment is available, rang- ing from simple, single-purpose devices to very elaborate, multi-purpose systems. We may again classify these according to the type of defect sensed, i.e., lump or hole detectors and optical detectors for spots, etc.

4.5.2 Lump detectors. These also detect other kinds of elevations, such as joins, folds, creases, etc. They operate by physical contact of the defect with the sensing head, which generates an electrical signal. There are two basic types, (a) those which are run in contact with the web and (b) those which are slightly backed off from the paper surface, only making contact with local elevations above a preset height. Type (a) gen- erates a continuous signal, which is the result of the dynamic response of the sensing head to random variations in paper thickness and from sheet flutter, etc. This type, therefore, has a signal-to-noise problem: on the other hand, it is capable of detecting small elevations, down to about 15 μm, under good conditions.

Type (b) receives no noise from the normal paper. However, backing-off from the surface sometimes presents problems, due to overall thickness changes as well as random, short-term variations. The practical minimum setting for this type is around 50 μm.

Some current methods are shown in Figure 3.

4.5.3 Join detectors. Paper webs are commonly joined by overlapping the ends. A join of this kind, therefore, extends the full width of the web and may be detected by a single, fixed- position unit. A join may have to be detected in one of several webs being cut at the same time.

Methods include capacity measurement, (the paper web or webs forming the dielectric of a capacitor) and photoelectric opacity measuring instruments, providing the total opacity is not too high.

4.5.4. Hole detectors. Mechanical hole detectors consist of metal brushes, which rest on the paper over a metal roll. A hole is signalled by contact of a bristle with the roll. Optical methods use a set of photocells on one side of the web, which detect light from a lamp on the other side of the web, when a hole is present. In some advanced designs, the signal-to-noise ratio is increased by using ultra-violet light: paper is more opaque in this region of the spectrum than in the visible band, also the effects of stray light are reduced.

Capacity methods are also used, for detecting relatively large holes and tear-outs. These are capable of detecting holes in one web of several being processed together.

4.5.5 Tear-out detectors. Tear-outs are the result of sampling from the edge of the web. To detect them, only single point inspection is required, located near the edge.

4.5.6 Combined-purpose units. Some inspection devices, e.g., capacity types, are capable of detecting joins, tear-outs and large holes.

4.5.7. Full-width optical inspection units. These are the most elaborate, sophisticated and costly instruments in the detection field. All operate on the same fundamental principle, the illumination of the sheet and the detection by photocells of reflected or transmitted light. A high performance is required, to distinguish small blemishes from the noise generated from random changes in the paper and from mechanical and electrical disturbances.

Two basically different approaches are used, (a) scanning and (b) multiple fixed photocell units.

The scanning systems inspect the paper by sweeping across the width of the web at a sufficiently high speed for successive scans to overlap. In the flying-spot method, a very small spot of light, about one mm diameter, is project- ed on to the sheet via a rotating mirror arrange- ment. (Figure 4). The latest versions of this instrument use lasers as the light source. A different system illuminates the paper with steady light and scans the image, another uses a combination of flying-spot and flying-image. The multiple photocell system (Figure 5) consists of a set of identical photoelectric units, each inspecting only a very small strip in the cross direction. The cells are sometimes connected in groups: paired units are used in some cases, with differential amplifiers, to balance out long-term variations.

Optical inspection units are capable of

detecting small faults, e.g., a black-on-white
spot of 0.4 mm diameter or less, performance depending on conditions, such as paper noise, web
stability, etc. Sensitivity depends upon optical
contrast. Complications arise where coloured paper
or coloured defects are involved, contrast depending on the colours of the incident light, the paper
and the defect and on the spectral sensitivity of
the detector.

The sensitivity of systems using transmitted
light also falls off with increasing paper opacity.

Optical units are generally not very efficient
in detecting lumps and creases and are often
supplemented by a lump detector.

Controls are provided to select accept/reject
levels. Some units are provided with facilities
to integrate the total number of faults in a sheet
or the total effect of low-contrast defects.

4.6 Applications

The optical or optical-cum-lump detectors may
be used at any appropriate stage of the process,
but most applications have been in the finishing
end to replace manual sorting. To do this required
the development of suitable mechanical handling
equipment.

In the earliest applications, machines were
developed for the sole purpose of replacing human
sorters, i.e., the machines accepted a stack of
cut sheets, which they separated into stacks of
accepts and rejects. Mechanical handling devices
lift the paper from the stack, one sheet at a time,
feed it through the inspection unit, thence to
separate conveyors, the separating device ("gate")
being actuated by the inspection unit. (Figure 6).

At the same time, it was recognised that if
the sorting operation could be combined with that
of cutting, a complete stage of the process could
be eliminated, with obvious cost savings, not only
in labour but also in handling damage. This required the development of special cutters, equipped
with double conveyor ("layboy") systems and a reject
gate.

The basic machine is shown in Figure 7. The
computer is required firstly to relate the position
of the fault detected in the web to the correct
sheet to be cut and, secondly, to delay actuation
of the reject gate until the leading edge of that
sheet is approaching it. Cutter-sorting is not so
simple as sheet-to-sheet sorting, because cutters
usually produce more than one sheet at a time. It
is common practice to cut several webs at the same
time (multi-reel cutting). It is also common
practice to sub-divide the web at the cutter, so
that each web cut produces several stacks of
sheets. This greatly affects the economics of
combining sorting with cutting.

The seriousness of the problem depends on two
factors, the number of sheets being cut at the
same time and the average reject rate. For example, if two webs are being cut together and each
is split into four sheets, a machine with a common
reject gate will reject eight sheets each time a
fault is detected. The reject stack will contain
seven-eighths good paper (ignoring coincident
faults). This will require retrieval sorting,

either by manual means or by a sheet-to-sheet
machine. This may be acceptable economically, if
the average reject rate is low, say 1%, but not
if it is, say 5%.

Possible solutions include the running of fewer
webs, e.g., single web and/or narrower webs, increasing the speed of the cutter to maintain
economical operation. Fewer webs may involve
running problems, especially with thin papers.
Another solution is to sub-divide the inspection
unit and the reject gate, in the cross-machine
direction, so that a fault detected at a particular position causes independent rejection only of
the corresponding sheet width. This solution
also introduces problems in runnability and complications, especially where the machine is to be
capable of dealing with a variety of sheet widths.
Few machines of this kind have been built.

The present-day trend appears to favour cutting single reel and sorting single width on relatively narrow machines, with high running speeds.

4.7 Performance

A large number of sorting machines is now in
operation. Some of these are based on "press-
stopper" rejection only, others employ full optical inspection. In general, results have been
satisfactory both in quality and in economic
return. In individual cases, however, experience
has varied widely. This could be expected, considering the differences that exist in sorting
requirements and in the types of inspection and
mechanical handling equipment, also considering
that the former had to develop from scratch and
the latter required development along new lines.

Some users found it possible to set up satisfactory accept/reject levels after only brief
evaluation. On the other hand, some applications
involved a very substantial amount of evaluation
and development.

A common problem in getting a project of this
kind off the ground is the basic difference between what the human sorter sees and what the
machine sees. One often finds that there exist
in the product a very large number of irregularities which beforehand were entirely unsuspected.
It is a feature of human observation of a routine
nature that one only sees the things one is looking for; the machine, however, sees defects in
terms of intensity levels. The machine is a much
simpler intelligence gatherer than the human;
it cannot differentiate between defects which
produce the same input signal but which differ
in appearance and seriousness. In practical
terms, a compromise solution is required.

The difference in discrimination has caused
a great deal of misunderstanding and, one sus-
pects, led to the abandoning of some projects
which could have been made to work. A prospect-
ive user would do well to examine his "good"
paper for acceptable irregularities just as
carefully as he selects those defects he wishes
to reject.

It has to be accepted that the machine will
not sort in quite the same way as the human; it
does not have the discriminatory powers. On the

other hand, it has the advantages of uniformity, reliability and speed. Judged on an overall basis of quality levels and customer complaints, it usually proves superior in performance and, in addition, has substantial economic advantages.

4.8 The systems approach

Automatic sorting is the most obvious application for inspection devices. However, to restrict consideration to this alone is to create an artificial barrier in the process. Defects originate in the raw materials or are generated within the process, each process stage producing its own kinds of faults. At each process stage, it may or may not be possible to stop generating defects, once their existence is known and there may or may not be an opportunity to remove faults, before the paper is transferred to the next process stage.

A defect costs money from the moment of birth. As it passes from one stage to another, its cost grows: it is wasteful to carry out processing on paper that already contains rejectable defects.

It is worth considering the problems of defects from a general systems point of view. For example, an inspection unit, installed at the end of the papermaking machine can have a twofold effect, feedback and feedforward. Firstly, signalling of defects may enable action to be taken to eliminate them at source, depending on the type of defect. Secondly, a record produced, showing the number of defects in each reel enables reels to be classified and decisions made for their disposal. Thirdly, defects can be identified by marking the edge of the web. This either enables them to be removed at one of the following stages or provides an easily-detectable signal for eventual rejection at a cutter sorter: in this case, the inspection unit on the cutter can be of a simple kind, to take out edge-marked paper, plus joins and any other simple faults introduced after the paper machine.

A scheme of this kind can be particularly valuable in the case of paper despatched in the form of reels, where the sorting process is impossible.

Extension of the use of inspection equipment to the papermaking machine and to other stages of the process is unlikely to have widespread use until the cost, compactness, ruggedness, reliability and simplicity of equipment make it an attractive proposition to the papermaker.

Looking into the future, we may see fully integrated units installed at different process stages, dealing not only with defects as defined herein but also with other out-of-tolerance properties such as weight per unit area, colour and so on. Such a scheme is technically possible at the present time, but may not yet be economically viable.

5 CONCLUSION

Electro-optical devices, currently in use in the papermaking process, perform various functions and can generally be regarded as useful and satisfactory.

The most prominent application is that of automatic inspection for defects. This has enabled manual sorting to be replaced by automatic means, special mechanical handling equipment having developed in step with the inspection units. Applications have been successful and profitable. However, uses are, so far, limited to papers of a sufficiently high quality to justify sorting. Any consideration of extension of applications to lower-grade papers must take into consideration (a) the problem of multi-reel cutting, (b) the fact that defects represent only a minor reason for rejecting paper and (c) that the economic returns cannot exceed the present cost of defects. Progress may lie in wider application of inspection units to other stages in the process or in integrating inspection for defects with instruments measuring other paper properties.

6 ACKNOWLEDGEMENT

The author wishes to thank the Directors of Wiggins Teape Ltd. for permission to publish this paper.

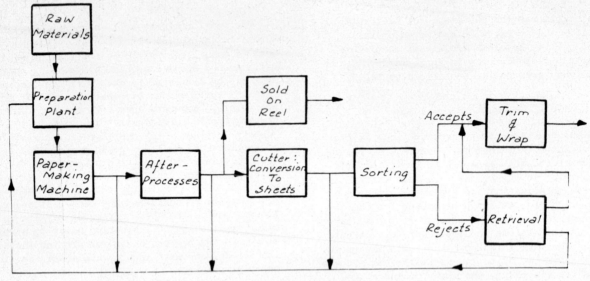

FIG. 1. BASIC PAPER MAKING SYSTEM

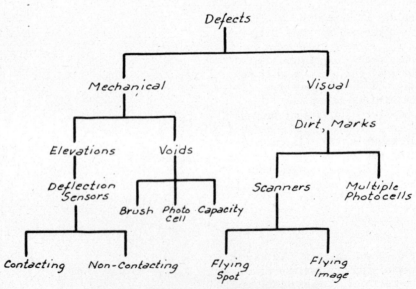

FIG. 2. INSPECTION SYSTEMS

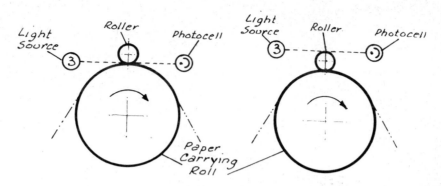

Contacting Methods, Using Light Beam

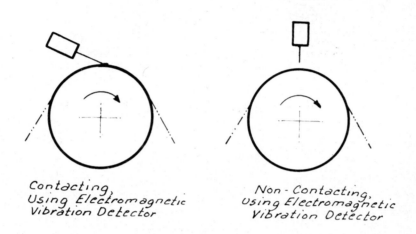

Contacting,
Using Electromagnetic
Vibration Detector

Non-Contacting,
Using Electromagnetic
Vibration Detector

FIG. 3. PRINCIPLES OF SOME TYPES OF LUMP
DETECTORS

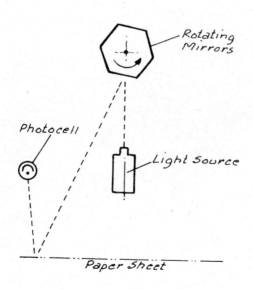

FIG. 4. PRINCIPLE OF SCANNER

55

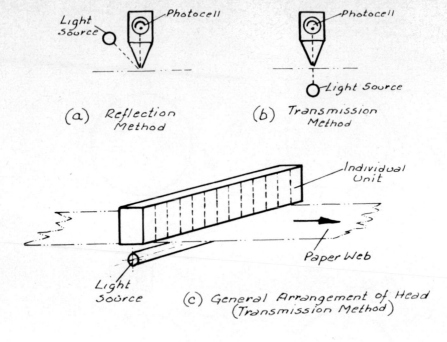

(a) Reflection Method (b) Transmission Method

(c) General Arrangement of Head (Transmission Method)

FIG.5. MUTIPLE PHOTOCELL SYSTEM

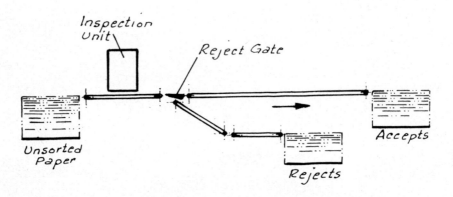

FIG.6. SHEET-TO-SHEET SORTER

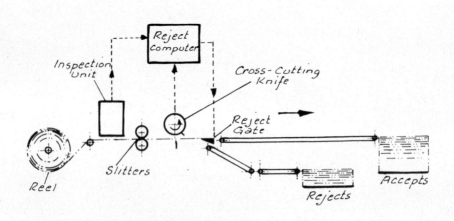

FIG.7. CUTTER-SORTER (SINGLE REEL CUTTING)

C128/74

SOME ELECTRO-OPTICAL EQUIPMENT AND SIGNAL PROCESSING TECHNIQUES IN THE FIELD OF MECHANICAL HANDLING AND SORTING

BASIL HARRY ROYSTON SPILLER, C Eng. MIEE, AMBIM
Principal Engineer, Decca Radar Limited,
Research Laboratory, Lyon Road, Walton-on-Thames, Surrey
The Ms. of this paper was received at the Institution on 10th December 1973 and accepted for publication on 31st January 1974. 44

SYNOPSIS   Application of the elementary straight beam photoelectric transducer to parts counting is considered.   A 'folded' light beam detector is described and some of the advantages resulting from its tolerance to the trajectory of parts through it considered.   Finally a multiple, parallel, channel counting technique is described which requires a minimum of parts organisation.

1.   INTRODUCTION

Few processes can be so tightly synchronized between input of material and output to the consumer that no buffer stock or bulk store of product is required. This paper outlines the development of an automatic sorting machine which will handle objects in the size range 0.5 mm to 100 mm.   Particular point is made of the adoption of a systems approach.

2.   THE SYSTEMS APPROACH

Fig.1. is a diagram of a possible system for a sorting machine consisting of a transport mechanism, a detector and a signal processor and controller.
It is probably safe to say that in general the mechanical sorting and transporting of objects presents the most difficult design problems.   This is particularly so when attempting to devise a versatile equipment capable of handling a wide variety of objects.   However, by adopting a system approach and considering the separate parts in turn and then together, a system was achieved in which the detector responded to quite imprecise delivery of parts resulting in a very simple and versatile feed system. The remainder of this paper outlines the techniques used in developing the Decca "Mastercount" system, commencing with the feed mechanism, moving on to the detector and finally some signal processing strategies.

3.   ORGANISING THE OBJECTS

3.1.   The Bowl Feeder
The bowl feeder, sometimes called a vibratory parts feeder, is a common industrial tool for sorting components, particularly when one component at a time is required for automatic assembly applications. When orientation need not be considered a high degree of organisation should not be necessary. Removing all but the essential tooling for single-file feed from the bowl feeder results in a versatile delivery system but parts leave the output launched on an unpredictable path if allowed to fall freely.

3.2.   The Belt Feeder
Typically this is a conveyor belt on to which a metered flow of parts to be sorted is introduced. A vertical, stationary wall sweeps the parts over to one edge of the belt so that all but a single row of parts fall off.   The retained objects leave the belt at the terminal roller.   Surprisingly, parts were found to have some degree of scatter and the following contributory reasons have been identified:-

1.   As each object negotiates the end roller, its centre of mass does not accelerate to match the angular velocity of the roller, the object becomes unstable.

11.   Some parts are slowed by contact with the stationary wall.

111.   Progressive contamination of the belt causes objects to adhere to it.

3.3.   Free Fall - As a Sorting Operation
Regardless of the type of feeder used, it is very likely a tight queue of objects will approach the feed exit before the detector stage and inter-object spacing may well be less than the optimum for detecting individual items.   Assuming velocity is much less than terminal and that parts are launched one at a time, then it is easily shown (Appendix 3) that the distance separating consecutive objects increases during fall.   (Time separation remains constant of course).   We can then produce a simple feeder to present objects to the detector with random time scatter but separated in space.   Using our systems approach it is now convenient to consider the detector.

4.   THE DETECTOR

4.1.   The Straight Beam Optical Transducer
Directly illuminating a photocell by a light source is an appropriate means of counting objects when they can be well enough organised to cause unambiguous modulation of the beam.   Where the geometry and reflective properties of objects to be detected allow, the light may be reflected from object to phototransducer.   Errors due to unwanted ambient illumination and similar disturbances are substantially eliminated if a modulated light source and appropriate signal filtering is used.   Solid state light emitting diodes are particularly convenient for use in this way.   Should levels of ambient light reaching the phototransducer be high enough to cause attenuation of the modulated signal, then no apparent change will be brought about by the

presence of an object in the beam. Little can be done except to restrict the acceptance angle into the phototransducer. Rejection of unwanted, fast signals and sometimes even those developed from debris mixed with the objects to be detected, can be achieved by integrating the amplifier's output due to the modulated light source to derive a voltage level. By choosing a suitable time constant for the integrator, disturbances brought about by a signal too short to be that from a 'legitimate' object will be rejected. Fig.2. is a block diagram of such a system.

Unambiguous detection starts with conjunction between object and light with the added condition that the change in light level brought about by the object's shadow must induce a photoelectric signal recognisable in the presence of the noise inherent in the phototransducer. Signals derived from objects of progressively smaller projected area with respect to the light beam's cross-section will ultimately become lost in thermally generated noise. It follows that reducing beam cross-section will improve the signal to noise ratio. Some difficulty can arise, however, through lack of positional stability of a small object in a narrow light beam, resulting in a series of signals being developed from a single object. One way of reducing these errors, which is also applicable to objects having holes in them, is to inhibit further response after the first for a period comparable to the object's normal transit time. Bearing in mind that the first signal might be developed somewhere near the trailing edge of some objects, it is clear that there is likely to be an upper limit on the optimum feed rate. Clearly a well-defined track ensuring that the object passes through the beam is a solution. However, this will demand a higher degree of parts organisation from the feed stage. In practice, such a scheme is likely to be difficult to achieve, not easy to maintain, be prone to jamming and result in a lower feed rate than that probable from a less organised delivery.

5. THE FOLDED BEAM DETECTOR

5.1. We have seen that for good signal to noise ratio in the optical detector, there is an upper limit to the beam cross-section in relation to the object size. This raised the difficulty of intercepting small objects with the correspondingly narrow beam, if it is intended to take advantage of the sorting action produced by free fall. The folded beam detector overcomes this problem. Fig.3 makes the principle clear. Also evident in Fig.3. is another advantage of this technique. The beam may be angled such that for much of the aperture between the mirrors an object up to half the beam width interrupts two segments and produces twice the modulation that would occur in a single beam. This characteristic is of great value in detecting flat objects since the smaller dimension cannot be in line with both beam segments.

5.2. Problems of Manufacture
Against the advantages to be got from folding up a long light beam was the apparent difficulty in achieving and then maintaining its precise setting. Fig.4. illustrates the effect of not having parallel mirrors and from such considerations it was decided to use a slightly divergent light beam of square cross-section. Divergence was found to compensate for mechanical errors and the square beam, produced by cylindrical lenses and mirrors on crossed axes to provide better coverage of the aperture.

Extensive use has been made of two sizes of folded beam detector, the smaller with a 50 mm x 50 mm aperture and a larger version has a 100 mm x 200 mm aperture. The smaller detector is intended to respond to objects of say 0.5 mm upwards and the larger detector, to objects from 3 mm typical dimension. The manufacturing tolerance derived from using a divergent beam of about 7° has meant that in their final form both detectors are basically 'C' shaped aluminium castings, that is with one side of the beam folding aperture open. Long mirrors of plate glass are cemented to machined surfaces and two adjustments only are required to align the reflections. The design has been found stable and entirely satisfactory for fitting to industrial systems.

5.4. Keeping the Detector's Mirrors Free from Dust
Virtually any photoelectric apparatus is vulnerable to dust coating on exposed glass surfaces and the mirrors of the folded beam detector perhaps more than most. The solution was found in inducing clean air, regulated to a gauge pressure between 2×10^{3} and 4×10^{3} Pa. (0.25 to 0.5 lbf. in^{-2} gauge), into the body of the detector to bleed from the slots over the mirrors. Higher air pressure was found to cause turbulence and deposit dust on the glass.

6. THE FOLDED BEAM DETECTOR IN A PARTS SORTING SYSTEM

6.1. We have now devised a detector which will accept parts delivered by the simple feed system discussed earlier. A folded beam detection placed below the lip of a bowl or belt feeder will detect parts delivered, the uncertain path of each part being within the aperture covered by the folded beam.

Obviously pulses developed from the detector's signals can be totalled on an electronic counter and control signals derived on reaching predetermined totals. (A pre-end of total signal can also be developed to slow the rate of parts delivery if necessary). An exacting application using a small bowl feeder (150 mm diameter bowl) and the 50 mm x 50 mm folded beam detector is given in Appendix 1 and Ref.1. Use of this type of detector, permitting parts to be in free fall, and hence their position after being detected to be predicted, means that a delay can be tolerated from detecting the last part of a batch quantity and deflecting the next part, when below the detector, to a new destination. This delay may be used to accommodate an electropneumatic solenoid valve's response and allow pneumatic cylinders to be used to operate substantial but fast operating diverting mechanisms. Adequate power has the added advantage of ensuring consistent operation in service.

7. SOME ADDITIONAL ELECTRONIC STRATEGIES

Having considered the feeder and detector system we will now discuss some electronic techniques which may be applied in principle to any system.

7.1. Synthetizing a Second Count Pulse
As has already been pointed out in 4.2. above, knowing how long modulation by an object should take can be used to reject spurious counts. The idea can, however, be extended to generate a second count pulse when modulation is too long for a single

object to have been the cause. The technique may sometimes help to reduce errors when pairs of parts are prone to becoming attached end-to-end or tangle. Fig.5. shows a block diagram of a pulse synthesizing arrangement and Appendix 2 shows a circuit based on NAND gates.

7.2. Rejecting Unwanted Transient Signals
In Section 4.1. it was suggested that fast, unwanted signals might be rejected by integrating. Where a non-modulated light source is used or for any other reason then signals too short to be from genuine objects may be rejected by comparing the signal's pulse length against that expected. If the signal is NOT still present after some time then no count pulse is developed. If, in Fig.5, the link between Monostable Pulse B into the counter pulse shaper is removed, this shaper will now produce a pulse only for (A.D.).

7.3. Adding the Output of Two Streams
Pulses from two detectors can be added on to one control counter. In this way manipulating parts may be made easier by halving the feed rate required. The technique is based on the Exclusive OR (sometimes called the wired OR) Gate.

The argument is based on there being a very small chance that two fast rise and decay pulses generated from separate detectors and amplifiers will have edge-to-edge coincidence. Fig.6. illustrates the principle.

8. AN EXPERIMENTAL MULTI-CHANNEL COUNTING SYSTEM

8.1. So far this paper has only considered parts organised into one output channel with the suggestion, in section 7.3 that two channels may be combined. Now if several channels were to be combined, it follows that having many alternative and equivalent ways of leaving the feed stage, the formal organisation of parts is less and in addition, the possibility of having some redundant channels might provide a system where jammed or tangled components blocking a channel is of little consequence.

8.2. One method of summing electronic count pulse signals from several channels in a single counter, irrespective of whether two or more counts occur simultaneously in different channels, is as follows:-

Each channel is provided with a detector and amplifier (see Fig.7.). The output from each amplifier (the data pulse) is fed to one input of a D-type bi-stable. A timing oscillator produces a train of pulses at frequency f_0, which is fed to a B.C.D. counter and thence to a decoder which has n outputs where n is the number of detector channels in the required system. In Fig.7. a 10-channel system is illustrated but for the sake of simplicity only channels 1 to 3 are drawn.

The decoder output pulse in any one channel is inverted and fed to one input of an AND gate, the other input being the output of the timing oscillator. The output of each channel gate is a succession of clock pulses, no clock pulses occurring simultaneously in different channels. For example, channel 1 clock pulses are triggered by oscillator pulses 1, 11, 21 .... channel 2 clock pulses by oscillator pulses 2, 12, 22 .... and so on. The object of the AND gate is to avoid

problems of delay time through the system by providing adequate time separation of the clock pulses between channels. The clock pulses in any one channel are fed to the other input of the D-type bi-stable associated with that channel. This bi-stable is set by the presence of both a channel clock pulse and a data pulse (C.D.), and is reset by the presence of a channel clock pulse and the absence of a channel data pulse (C.D.). The set operation of the bi-stable produces an output which is fed via a suitable coupling circuit to the counter input. Since no clock pulses occur simultaneously in different channels, no two count pulses can ever be fed simultaneously to the counter. The first channel clock pulse to arrive at the bi-stable after the onset of a data pulse sets the bi-stable and produces a count; the arrival of the first channel clock pulse after the termination of a data pulse resets the bi-stable ready for the next data pulse in that channel.

With an oscillator frequency of f_0, the maximum count rate in any one channel is given by $f_0/2n$, where n is the number of channels. In cases where the mark/space ratio of the data channel is high (e.g. where there are short intervals between long data pulses, as may occur when long objects are being fed nose-to-tail), the count rate in a single channel can be increased without raising the value of f_0 by the following method:-

After the termination of the data pulse, instead of waiting for the next clock pulse in the particular channel to reset the bi-stable, the next oscillator pulse can be used. This is achieved by employing a three input gate whose inputs are:-

    (1) The state of the bi-stable (set or reset).
    (11) The presence or absence of the data pulse.
(111) The oscillator pulses.

This gate is used to produce a reset signal under the following conditions:-

(a) bi-stable set
(b) data pulse absent
(c) oscillator pulse present

With this method, the maximum channel count frequency can be nearly doubled, to approach f_0/n.

If T is the minimum data pulse duration f_0/n must be greater than $1/T$ to ensure that at least one channel clock pulse arrives during each data pulse. It is immaterial if the oscillator frequency f_0 is greater than the above-stated minimum, since once the bi-stable has been SET, the arrival of further clock pulses will have no effect until after the termination of that data pulse. Since with an oscillator frequency of f_0, two successive count pulses can reach the counter at an instantaneous rate of f_0, the counter cut-off frequency must be at least f_0.

9. CONCLUSIONS

In applying electro-optical detectors to mechanical handling and sorting, a systems approach should be used. Optimising the method of detecting the part in isolation from the apparatus to be used to sort and present the objects to it is likely to produce a detector of less than optimum systems effectiveness.

Though of undisputed general usefulness, the simple

straight beam detector puts unreasonable demands on
the performance of the sorting and feed system,
particularly when small objects have to be counted
and batched from a chaotic collection.   A folded
beam detector has been suggested which offers some
relief from the difficulties of monitoring the output from typical sorting and feeding apparatus
unable easily, to provide a well-defined pathway
for the output stream.

Some suggestions have been made of how signal processing might be of help in reducing error counts
in signals from optical detectors.

Finally signal processing in an experimental counting system has been described able to accept random
parallel count pulses.   Such a system might find
application in assembling very large batches of
objects fed at a high total through-put rate or
count batches of very difficult-to-handle objects
by many parallel, slow delivery channels or it might
perhaps perform the task of an accurate filter
metering difficult-to-handle objects into a more
conventional batching stage.

10.   ACKNOWLEDGEMENTS

The author wishes to thank Mr. S. R. Tanner, B.Sc.,
C.Eng., F.I.E.E., F.Inst.P., A.K.C., Research
Director of Decca Radar Limited for permission to
publish this paper.

An Automatic Device for Counting Dry Fish Eggs

Using a small counting system based on the 50 mm x 50 mm folded beam optical detector described in the paper, and a small bowl feeder of approximately 150 mm diameter bowl, Bayer and Clifford have reported (Ref. 1) the following performance achieved by the apparatus when used to count dry eggs of herring, average egg diameter 0.9 mm and trout, average egg diameter 5.0 mm.

Table 1

Counts of known numbers of fish eggs made at various feed rates.

| Test | Herring | | | | | | Trout |
| | 2 000 eggs | | | 50 000 eggs | | | 1 700 eggs |
Number	20/sec.	30/sec.	35/sec.	20/sec.	30/sec.	35/sec.	20/sec.
1	1 998	1 972	1 904	49 837	48 941	48 194	1 698
2	2 000	1 968	1 902	49 700	48 770	47 831	1 697
3	1 996	1 972	1 901	49 800	48 835	47 387	1 699
4	1 997	1 973	1 915	49 777	48 697	47 521	1 700
5	1 998	1 966	1 922	49 632	48 915	47 441	1 699
6	1 997	1 971	1 918	49 841	48 850	47 477	1 700
7	1 996	1 964	1 910	49 715	48 650	48 141	1 700
8	1 997	1 975	1 922	49 634	48 552	48 005	1 698
9	1 998	1 969	1 918	49 652	48 602	47 421	1 700
10	1 998	1 970	1 928	49 857	48 700	47 620	1 700
Mean	1 997.5	1 970.0	1 914.0	49 744.5	48 751.4	47 713.8	1 699.1
% Error	0.02	1.5	4.3	0.51	2.5	4.6	0.05
Standard Deviation	1.2	3.3	9.33	88.78	132.38	295.21	1.1

APPENDIX 2

Fig. 8 below shows one circuit based on NAND gates, specifically the common 74 Series Transistor to Transistor Logic (TTL) family.

Three dual in-line 7400 packages, each containing four, two input NAND gates were used for the prototype circuit. There are no doubt numerous, probably more elegant, alternative designs based on current integrated circuits.

Fig.8. Circuit Operation

Signal A Logical 1 is applied to gate 1 input and to the counter. A count results. All gates are NAND hence we have an output of NOT 1 for 1 input. So Gate 1 output - Logical 0 into Gate 2 gives output of Logical 1 into Gate 3 to give Logical 0 output through C into transistor Tr. 1 base. Tr.1 turns off for a period 0.8CR (approximately). (C in Farads, R in Ohms). So Gate 4 input is Logical 1 for 0.8CR then its output is Logical 0 for 0.8CR which 'clamps' one of Gate 2 inputs to Logical 0 for this period. This monostable action is repeated in the second pulse shaper. Gate 4 Logical 0 into Gate 5 makes Gate 5 output Logical 1 for the period set in the first monostable but on recovery Gate 5 output goes to Logical 0 triggering the second monostable and Gate 9 output to Logical 1, the C pulse.

If then A, the transducer developed signal, is still present then A.C develops a second count pulse. (Gate 11 inverts A.C to get a positive pulse).

If pulse A is not taken to the counter, but only the pulse due to A.C then the circuit behaves as mentioned in Section 7.2 above.

APPENDIX 3

If distance S_1 is traversed by the part falling for t_1 s.

Then $S_1 = \frac{1}{2} g t_1^2$

and S_2 for the next part falling for t_2 s.

$S_2 = \frac{1}{2} g t_2^2 \qquad t_1 > t_2$

and vertical separation is:-

$S_1 - S_2 = \frac{1}{2} g (t_1^2 - t_2^2)$

If $nt_1 = t_2$ where $1 > n > o$

$S_1 - S_2 = \frac{1}{2} g t_1^2 (1 - n^2)$

and $\frac{d}{dt} (S_1 - S_2) = g t_1 (1 - n^2)$

that is the rate of change of separation with time of fall is a constant and always positive therefore vertical separation will increase with time.

REFERENCES

1. BOYAR H. C. and CLIFFORD R. A. Transactions of the American Fisheries Society. Vol.96, No. 3 21 July 1967 pp. 361-363.

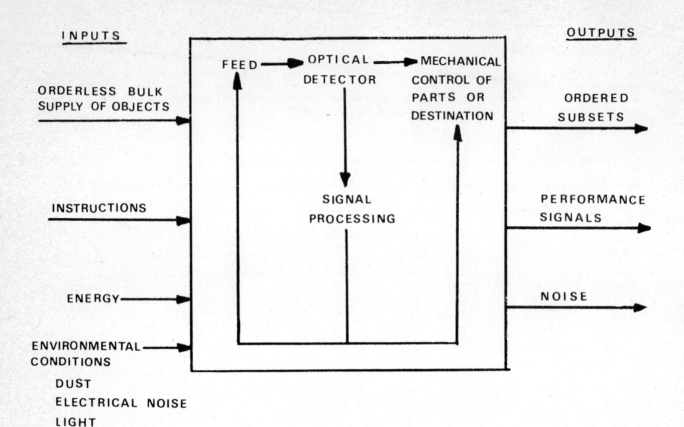

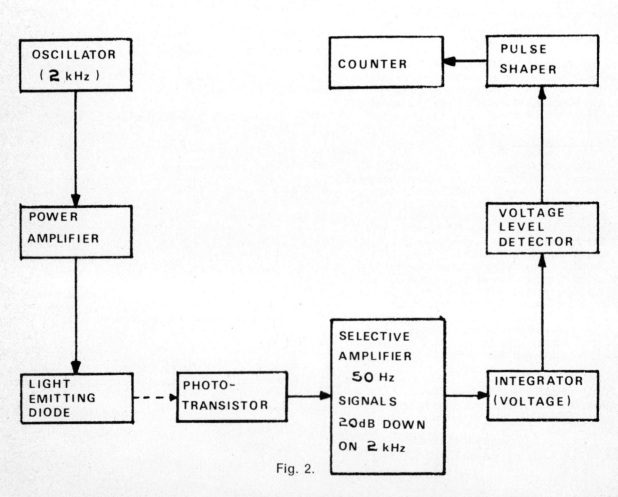

Fig. 1.

Fig. 2.

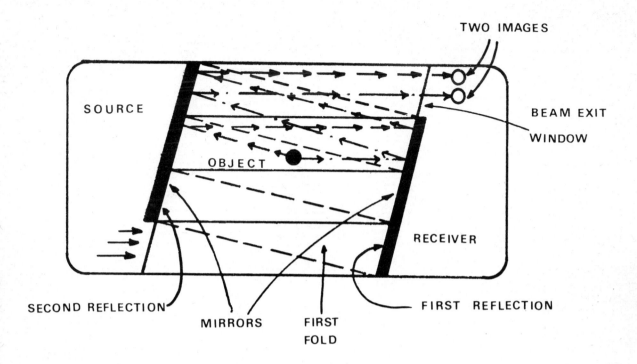

Fig. 3.

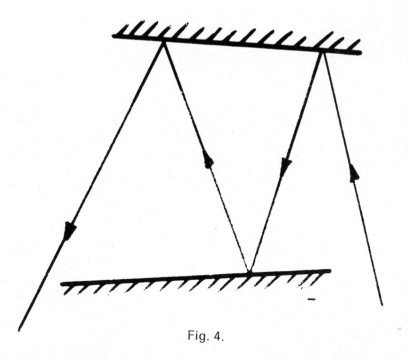

Fig. 4.

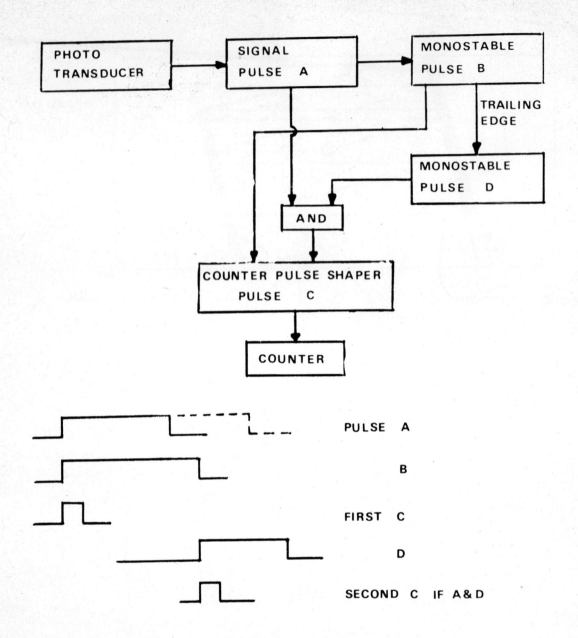

Fig. 5.

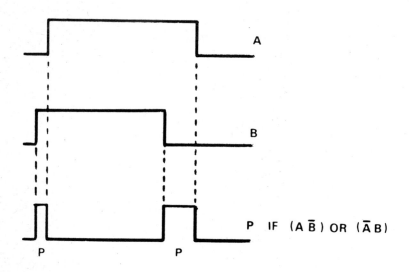

A	B	C
O	O	O
O	I	I
I	O	I
I	I	O

A

B

P IF $(A\bar{B})$ OR $(\bar{A}B)$

P   P

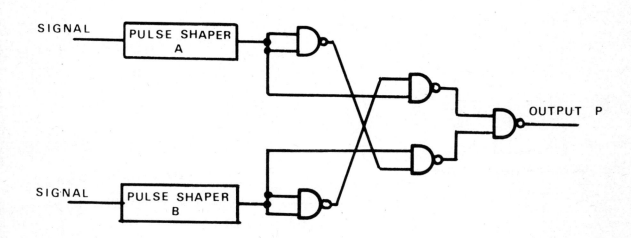

Fig. 6.

65

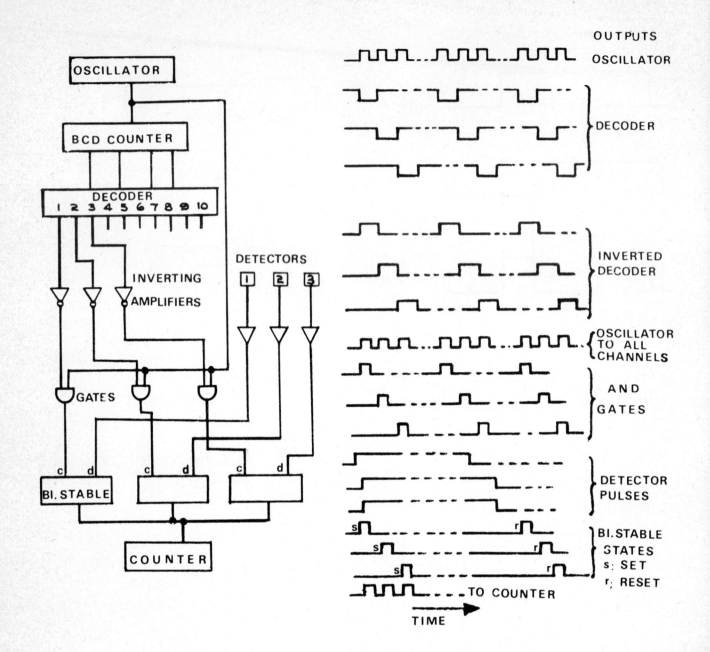

Fig. 7.

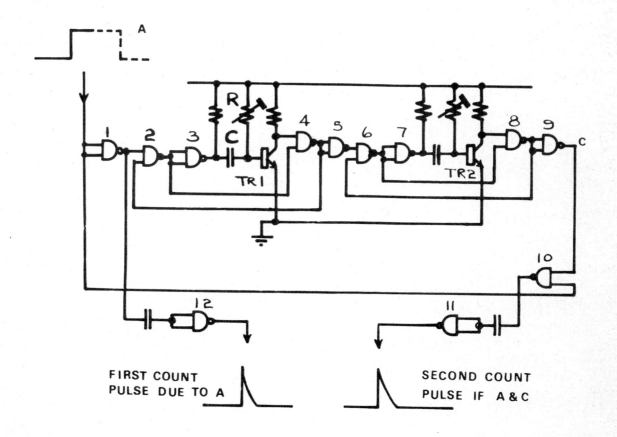

Fig. 8.

OPTO-ELECTRONIC COMPONENTS FOR USE IN MECHANICAL HANDLING AND SORTING EQUIPMENT

P. TURTON
Marketing Manager, The Plessey Co. Limited,
Wood Burcote Way, Towcester, Northants NN12 7JN

1. OPTOELECTRONIC DETECTORS

The title detectors applies in the context of this paper to devices whose electrical characteristics are changed when they are exposed to visible or near infra-red radiation. These detectors are grouped as follows, according to their mode of operation: photoconductive, photovoltaic, phototransistors and integrating detectors.

1.1 MATERIALS

Many semiconductor materials are available for the manufacture of detectors but practical limitations reduce the choice to cadmium sulphide, cadmium selenide, germanium, silicon and selenium (CdS, CdSe, Ge, Si and Se). The last named material is used mainly for single photovoltaic cells for the measurement of light levels. The other four are used for the manufacture of both single and complex arrays of detectors for industrial applications.

Detectors manufactured from cadmium sulphide and cadmium selenide are in the photoconductive group since their conductivity increases with increase in incident radiation. Detectors manufactured from germanium are almost all phototransistor whilst those using silicon as the starting material can be manufactured to be photovoltaic (photodiodes), phototransistors and integrating light detectors, whose operation will be described later.

1.2 MANUFACTURING PROCESSES

Photoconductive devices can be made in large areas and are usually manufactured by spraying an aqueous suspension of high purity CdS or CdSe plus dopants and binders onto a ceramic substrate. This is dried and then sintered in a controlled atmosphere to maintain the required dopant levels and hence control the photosensitive properties of the device. Electrodes of an interdigitated form are then applied by spraying or the evaporation of metals through a stencil type mask. The interdigitated electrode pattern allows what would otherwise be a long but narrow gap to be condensed onto the square or circular substrate to give a photocell of large area, say up to 1.5" square. The electrode pattern can of course be divided into many sections to provide an array of separate cells on a single substrate. Glass envelopes, plastic cases or metal envelopes with glass windows are used to encapsulate photoconductive devices.

Germanium detectors in the area of interest are largely phototransistors manufactured on the alloy junction process. In this process a small slab of n type germanium has two small pellets of indium, alloyed in to form two pn junctions, one pellet on each side of the slab giving a pnp structure. One pellet acts as the emitter and the other pellet as the collector of the transistor. The Ge slab in the immediate vicinity of the pellet is light sensitive and therefore the device can be used as a phototransistor with the slab acting as the base electrode. Small glass envelopes with hermetic seals are used to encapsulate this type of phototransistor.

The remaining material on our list, silicon, is probably the most widely used of all semiconductor materials. It is used for photodiodes, both simple diffused and planar diffused, phototransistors and integrating light detectors (ILDs). Photodiodes are photovoltaic, that is they generate electrical power from incident radiation. This electrical power can be detected as a voltage or current depending on the application and electronic circuitry employed. A silicon phototransistor is a structure having an enlarged base collector junction which is sensitive to light. Current generated in this region is an internally generated base current and is therefore amplified when injected into the base emitter junction of the transistor. (See Fig. 1). Higher gain can be obtained by the inclusion of a second transistor connected to the first in the Darlington configuration. The third category of silicon device is the ILD based on the M.O.S. integrated circuit process using photodiodes, M.O.S. transistors and resistors. As its name implies, it successively integrates the total light falling on the detector during predetermined periods giving an output proportional to that light level.

All modern silicon devices are manufactured on high temperature diffusion processes where various dopants (or controlled impurities) are introduced into the silicon to alter its electrical properties. In this process polished silicon slices in the range of 0.007 to 0.015 inches thick, which are initially doped to have n type conductivity, are placed in a furnace at a high temperature. The atmosphere and temperature of this furnace can be controlled accurately in the region $900^{\circ}C$ to $1000^{\circ}C$.

A carrier gas, containing a proportion of a vapour of a volatile source of boron such as boron trichloride is introduced into the furnace. The boron trichloride decomposes at the furnace temperature and deposits boron onto the silicon. This boron then diffuses into the silicon crystal lattice and converts a thin surface skin from n type conductivity to p type to form a pn junction. The areas onto which boron is deposited and hence the areas where pn junctions are formed can be defined by the use of diffusion masks formed by photoengraving suitable "windows" in a previously formed skin of silicon dioxide or nitride.

In a semiconductor material of n type conductivity the electric current is a flow of excess electrons whereas in p type conductivity the current is a flow of holes (or positive charges).

The pn junction formed by the first diffusion is employed as the photosensitive region in the various devices. Devices so produced are called planar diffused devices, either photodiodes, phototransistors or self scanned arrays. If a simple photodiode is required then contacts are made to the n and p regions by vacuum deposited aluminium films.

If the device is destined to be a phototransistor then a second diffusion is required. This is a small area diffusion to convert part of the p type region back into n type conductivity to create an emitter for the transistor. Arsenic or phosphorous are used as dopants for this operation where the desired areas are defined again by a mask formed from an oxide or nitride skin.

Contacts to the three regions, npn are made by a vacuum deposited aluminium film photoengraved into the required pattern.

A simplified version of the photodiode process is used widely to produce cheaper photocells for industrial uses where electrical specifications can be relaxed. In this process the p type skin is allowed to form on all surfaces of the silicon slice. This skin is later removed from the back of the slice to permit connection to be made to the original n type material. Contacts are made to both p and n type regions by nickel plating which is deposited through a stencil type mask. The plated areas are then solder coated and the slice is broken into individual photocells of the required size.

The M.O.S. or more correctly field effect transistor (FET) process, used to manufacture ILDs, has only one diffusion operation. This is of p type conductivity into an n type starting material and the p type regions are arranged to define photodiodes, resistors or transistors. The transistor is formed from two closely spaced p regions (source and drain) with a metal field electrode (gate) controlling the conductivity of the gap between the two diffusions. Again an interconnection layer is provided by a vacuum deposited metal film photoengraved to the required pattern.

The photodiode and phototransistor are referred to as discrete devices, i.e., one detector per package, whereas the ILD is an integrated circuit whose complexity varies according to its function and can have many detector elements in one package. A single ILD may have only 20 components but the very complex optical pattern recognition devices have in excess of 3000 components consisting of detector diodes, resistors and transistors.

2.1 MECHANICAL CONFIGURATIONS OF DETECTORS

Photoconductive devices have sensitive areas varying from a few square mm, say 2mm x 1mm, to ten of square mm, say 400mm x 1mm or 900mm x 0.25mm (the smaller dimension is the spacing between the electrodes and the larger dimension is the electrode width). The glass or plastic encapsulations are generally large, up to 32mm diameter and 32mm long. This size often restricts their use to single cell

installations for counting objects. Although it is possible to form multiple cell arrays these are uncommon and generally outperformed by silicon devices of smaller area. The glass envelopes result in a fragile device not suited for use in applications where high levels of vibration or shock are experienced.

The diffused silicon photodiode can have active areas from say 0.8mm x 2mm up to 20mm square. These cells can be provided with wires or soldered either singly or in groups onto printed circuit boards for support.

The positioning of the photocells on the PCB can be tailored to meet individual requirements of the handling or sorting equipment. In this case the spacing between photocells can be large and they need not be in a single row. Circular, rectangular and angled arrangements are all possible. When the application demands many photocells in a single row with spacings less than 1mm, multiple cell arrays can be obtained from a large single photocell. The active layer of the large photocell is divided into smaller sections by etching 0.5mm to 0.25mm wide grooves through this upper layer until the sections are electrically isolated. By this method the large photocell can be divided into about 15 sections with each section having a separate top connection. This multiple cell chip can then be soldered directly onto a PCB with all the connections being brought out separately. Connector pins or flying leads can be provided on the assembly to facilitate connection into the equipment in which the array is used.

Changes in dimensions of the photocells and arrays are simple since the machines used for producing the etching masks can be programmed to move in steps of 0.001 inch. Hence pitch or cell lengths can be varied in steps of 0.001 inch giving great flexibility in their applications.

The planar diffused process can also be used to fabricate single or arrays of photocells. By this method it is possible to have diodes arranged in rectangular matrices or single rows. Because the processing masks are made using photolithographic techniques the diode dimensions and spacings can be extremely small. With modern technology it is possible to make diodes only $25 \mu m$ square with $12 \mu m$ spacing between adjacent diodes but for practical applications they are commonly in the range $50 \mu m$ square to $250 \mu m$ square. In certain cases more area is occupied by connections than by the diodes themselves.

Several manufacturers, mainly in England and the U.S.A. have taken the opportunity to improve performance and reduce overall equipment costs by modifying the simple planar diffused photocell array to include amplifiers and other active components on the same piece of silicon.

As previously described the planar photodiodes use the same diffusion of the M.O.S. process as is also used for transistors and resistors. Therefore it is possible to condense a large part of the electronics required with photodiode arrays and make light detecting integrated circuits (also known as self scanned optoelectronic detector arrays) by connecting the individual components in the desired circuit pattern. This pattern is determined by the intended application of the detector array. Linear arrays are used for optical character recognition and facimile readers whereas rectangular arrays are used for imaging and pattern recognition.

As with the simple planar photodiode array mentioned above the diodes are in the range 50 μm to 250 μm square. The photodiodes however only occupy between 50% and 70% of the device chip area. The remainder is occupied by the associated amplifier and switching circuitry giving an overall chip size in the range 1mm square to about 12mm x 3mm. The larger chips have in excess of 3000 individual components interconnected by wiring pattern photoengraved from a film of vacuum deposited aluminium. It is possible with linear detector arrays to construct them using the "chocolate block" principle where the detector elements and their switching circuits are repeated in rows across the silicon slice and after processing are separated into chips containing any desired number of elements.

Reduction in tooling and design costs are a result of this approach.

With only one exception all the devices in the category of self scanned optoelectronic arrays operate in a serial mode, i.e., the sequential interrogation of each detector element is controlled by inbuilt electronic switches. The exception is a Plessey device having a 72 x 5 format intended for O.C.R. applications where serial scanning is used in the 72 direction down each column and each of the 5 columns has a separate output giving serial parallel operation. Serial scanning is almost a prerequisite for these devices since it would be impractical to provide separate connections for each detector element if parallel outputs were required. For a small array of only 100 elements on a 2mm square chip over 100 connections would be required but these could be spaced at 300 μm (0.006") centres on the chip. Unfortunately it is not possible to keep to this spacing in the device package and the PCB design which accepts the package. A more usual lead spacing at the package level is 1.25mm making the minimum package periphery 125mm i.e., about 11mm square, plus an allowance for leads of a further 3 or 4mm. The overall area occupied by this detector would be slightly less than 2cm square which is not too formidable even when making over 100 soldered joints in this space. The connections to large arrays of say 250 and 500 detector elements would however present a formidable task and occupy areas of about 75mm and 100mm square respectively.

The serial mode of operation offers a solution to the problem by only requiring less than ten output connections irrespective of the array size and permitting a small familiar dual in line package to be used for encapsulation.

Phototransistors, because of their inbuilt amplification can be relatively small with active detection areas in silicon types of between 0.4mm and 1mm square. These small chip dimensions permit small packages to be used, the smallest being the "pill" package. This has an outside diameter of 1.5mm and length of 2.3mm and is suitable for assembly into high density arrays on PC boards. Larger packages such as TO5 and TO18 windowed cans, taken from the electronics industry standard TO range of packages, are used where high packing densities are not required.

Some specialised pattern recognition assemblies have taken advantage of recent developments such as beam lead and hybrid assembly techniques to fabricate arrays having up to 120 phototransistor chips on 1.5mm centres. The beam lead assembly process uses specially prepared silicon dice where, during the final stages of manufacture, the connecting pads are extended by vacuum evaporation and electroplating to form "beams" which extend in cantilever fashion over the edges of the dice. It should be noted that the "beam" is not the Mechanical Engineers usual concept of a beam since it is only about 0.25mm long, 0.1mm wide and 0.025mm thick. The chip relies, during assembly, on the support of these beams. Nevertheless machines have been developed both in England and the U.S.A. for assembling such chips directly onto metallised ceramic substrates for the fabrication of complex circuits. All the beam connections on each chip are made simultaneously by thermocompression or ultrasonic welding.

In the hybrid assembly process standard unencapsulated but passivated photo-transistor chips are attached to metallised ceramic substrates by special gold alloy solders or conducting epoxy pastes. The connections to the chip being made again by thermocompression or ultrasonic welding. Both types of assembly are provided with connecting pins fixed into the ceramic and are completed by being encapsulated in a clear plastic resin.

3. COMPARISON OF DEVICES

The large sensitive areas of photoconductive devices made them suitable for applications where high currents and slow switching speeds are required and low accuracy switching points can be tolerated. Applications such as counting objects on conveyor belts or detecting when doors are open or closed are typical. They were the first practical semiconductor replacement for the vacuum photocell. These devices can control currents of up to 30mA and voltages in the range 10 to 50V and hence can operate small electromagnetic relays without any intermediate electronics. Operating speeds must be restricted to a few operations per second and may require movements of 10mm to effect the switching levels.

The cell resistance can be reduced from its dark value of 2 to 5 MΩ down to 1KΩ or less by the application of a light of about 1mW/cm^2 intensity. The spectral response is narrow and in the visible (see Fig 2) with CdS having a peak response at 550nm and CdSe at about 650nm. It is difficult therefore to use them in conjunction with solid state emitters since visible LED's cannot emit sufficient power to be useful.

In spite of many shortcomings photoconductive cells are cheap and permit low cost installations to be obtained.

The simple diffused photodiode is the most versatile device of those being considered here. It can take the form of single photocells varying in size from 0.6mm x 2mm up to 20mm square. In multiple cell chip form it varies from 2 to 15 cells per chip with a minimum 2 cell chip size of 2mm square. A wide spectral response from 450nm to 1100nm makes the devices suitable for operation with tungsten filament lamps or solid state infra red emitters. The peak response of the silicon photocell is at 900nm and coincides with the emission wavelengths of 900 to 950nm from gallium arsenide infra red L.E.D.s making all-solid-state detector/emitter units practical. Three modes of operation are possible with the photodiode. The first mode of operation, in which the short circuit current is measured, gives a linear increase in output current for an increase in incident radiation. When the external load resistance is low linear operation over 5 orders of magnitude can be achieved. At low intensities the output current may be only a few microamps, but at high intensities a few milli-amps may be generated. In this mode the device is very stable with changes in temperature, only about 5% change in current occurring for a change of say 40° in cell temperature. Also the effects of production spreads are least noticeable on short circuit current operation. Because of this it is possible to make the 10 to 15 cell arrays with photocurrents matched to ±10% of the mean. Both phototransistors and photoconductive cells have sensitivity spreads of 3 to 1 which requires the associated amplifiers to have a wider range of adjustment to accommodate these variations in detector sensitivity. For all practical purposes no speed limit is imposed on mechanical equipment by the photodiode since it will respond to $1\mu S$ pulses of light.

The second mode of operation is to measure the current through the photodiode when it is reverse biased. The performance is approximately the same as when using the short circuit current mode but it is more temperature sensitive. Faster response times can be achieved because of the reduction of cell capacitance with applied voltage.

Thirdly if linearity is not required then the open circuit voltage mode of operation can be used. This mode of operation has the advantage of giving higher voltages at low light levels since the voltage is a logarithmic function of incident intensity.

Direct comparison of basic sensitivity between devices in the different groups is difficult since the effect of light produces different changes in various devices. A figure useful in making comparisons of sensitivity, particularly for silicon devices, is the minimum incident power detectable above the device noise. For simple diffused photodiodes the level is about 0.005mW/sq.cm; for planar diffused photocells and phototransistors the level is 0.001mW/sq.cm. but for I.L.D's the level is 10^{-10}mW/sq.cm. or less. These levels are not necessarily related to the device output at high light intensities. Measurable changes from the dark resistance value of a photoconductive device can be obtained with a light level of about 0.01mW/cm².

Since external amplification is necessary with photodiodes and phototransistors the choice of these devices in a large number of applications is the personal preference of the design engineer or the effectiveness of the respective salesman. Approximately the same overall electrical performance can be obtained from either device. Mechanical requirements for close spacing or environmental stability can override electrical performance comparisons.

4. COMPARISONS OF COSTS

As the complexity of the detector increases so do most of the costs associated with its development, production and use. Development and tooling costs for simple silicon photocell arrays are usually so low as to be insignificant. The PCB used for mounting the array will cost about £30 to £50 for the preparation of artwork and simple production jigs. With phototransistors it must be assumed that a small range of standard chip specifications are available. Assemblies of standard packages onto PCB will require the same artwork and jigging as the photocell and costs £30 to £50. When assemblies of phototransistors or single I.L.D's into special packages are carried out the costs rise rapidly.

Artwork and printing screens for the connection pattern must be made but also assembly machines may require modification. Proving the assembly and final test equipment can give a bill totalling £500 to £2000.

The costs of production silicon devices varies from about 50p to £1.50 per detector element considered above.

The special I.L.D. used for OCR and pattern recognition can offer tremendous improvements in machine performance and size but its development costs are also high. A bill for £20,000 must be anticipated for the design and manufacture of a 200 to 500 element array. A single row of say 25 detectors could be obtained for about £6,000. These costs are built up from circuit design time, test circuit manufacture, artwork and mask preparation and final test equipment. Final production devices will cost in the range of £40 to £120 excluding any contribution from the development costs which are usually set against the system development. Although these I.L.D. costs are high the equipment is viable since it is the only solution available to efficient character recognition.

5. SOLID STATE LIGHT SOURCES (L.E.D.s)

The present status of light emitting diodes as light sources in industry is considerably lower than that of the tungsten filament lamp. Several factors contribute to this state of affairs but the main two are high efficiency and absolute optical power obtainable from the lamp. With the continued development of gallium arsenide in particular, the small filament lamps, of below say 3 watts input, are being seriously challenged in all industrial applications.

Whilst filament lamps are made with power inputs from say 0.25 W to 1000 W, L.E.D's either infra red or visible are measured by their output optical power and range from 0.5mW to 200mW under DC conditions. Most infra red emitters can be pulsed at 10 to 30 times their DC maximum current with a corresponding increase in power output. It is sometimes advantageous to operate the L.E.D's under pulse (or effectively AC) conditions and arrange the detector to be followed by an AC amplifier. The system can then be made immune to DC changes in light level caused by sunlight or other ambient lighting falling on the detector.

The light emitting diode like the photodiode and transistor relies on a pn junction for its operation. The junction fabrication process is similar to the photodiode process where controlled impurities are diffused into the base material to give layers of different conductivity. When the junction is forward biased electron hole recombination occurs at implanted impurity centres and light is emitted. The wavelength of the light is determined by the band gap, either direct or indirect, of the semiconductor material.

Gallium phosphide and gallium arsenide are the two basic materials used for L.E.D's although the GaAs can be modified to gallium aluminium arsenide and gallium arsenide phosphide. Fig. 3 gives the range of emitted wavelengths of L.E.D. made from these materials, and Table 1 gives the range of output powers and operating currents.

It will be seen that solid state light sources are fairly monochromatic and can be used without the use of filters to give lights of different colours or infra red radiation from 550nm for GaP green to 950nm for GaAs infra red. The visible light emitters gallium phosphide and gallium arsenide phosphide are used almost exclusively for panel indicators, warning lamps and numeric indicators. Gallium arsenide (GaAs) and gallium aluminium arsenide infra red emitters are used in industrial applications to replace filament lamps for illuminating photocells. Silicon detectors and GaAs emitters are admirably suited since their peaks of sensitivity and emission coincide. The long life of operation obtainable from an L.E.D. is its major advantage over filament lamps. This long life can be maintained even when subjected to high vibrations and shock. These two conditions are always encountered in any mechanism however small or well balanced and are the worst enemies of filament lamps.

Infra red L.E.D's are available with emitted optical powers in excess of 20mW at 300mA with a voltage across the diode of 1.5 Volt, 450mW then being dissipated. A better performance can now be obtained from the infra red L.E.D. because only about 7% of the input wattage of a filament lamp is converted to light, i.e., about 32mW output for 450mW input. Whereas the 32mW is radiated in all directions from the lamp the 20mW of radiation from the L.E.D. is within a cone of about 60° and can be collected by a simple lens. Therefore unless a well designed reflector is used with the filament lamp only a few mW of optical power is available for use. Nothing unfortunately is free - the 20mW output L.E.D. costs about £11 in small quantities but the $\frac{1}{2}$ watt filament lamp of high quality may cost up to £2. Since L.E.D. life is several times that of the lamp, well in excess of 100,000 hours, the initial cost of its installation is usually saved by the elimination of only one service call to replace a failed filament lamp.

The majority of applications of infra red L.E.D's use only a single emitter but use some form of optical system to distribute the radiation to the desired points opposite arrays of photo detectors. It is however possible to use the hybrid assembly techniques as used with the silicon detectors with L.E.D. chips and arrange them in any desired pattern. Commercial assemblies of L.E.D's are available which align with the arrays of detectors used in tape, card and badge readers. Selection of the L.E.D. chips is necessary to obtain equal emission from all points.

The L.E.D. is well suited to the user of numerically controlled machines, tape readers, card and badge readers, all safety interlocks on machines and, in fact, any application where light is used to detect the presence of an object and indicate its presence to an operator. They cannot at present help the people engaged in coding and sorting parcels and letters at the G.P.O., where a phosphor dot code is printed on the object. This phosphor is activated by ultra violet light and emits visible radiation which can be sensed by any of the detectors discussed in the first part of this article. Unfortunately no L.E.D's are available which emit either ultra violet or blue light and although it is known that some materials can emit blue light it will be many years before blue emitters are commercially available to activate the phosphors used in letter coding.

It is often required to display alpha numeric information at some point in a handling system. The information can be of distance of machine move-ment along its axis, weight, set dimensions or number of operations performed or objects counted. Again processes developed for producing silicon integrated circuits and hybrid assemblies have been used for the production of L.E.D. alpha numeric indicators. These are available in the three standard L.E.D. colours namely red, yellow and green and range in character height from 0.1" high to 0.75" high. Packages containing only one digit or up to 16 of the 0.1" high version are produced. The multiple character packages reduce packaging costs and allow a more aesthetically pleasing presentation of the information with the 0.1" high characters.

All the usual advantages of L.E.D's such as long life apply to the digit indicator. An extra advantage over say neon digit indicators is the ability to produce L.E.D's in three colours and thus ease problems of reading displays with many rows of characters.

6. APPLICATIONS

One mechanised device which has benefited greatly from the use of optoelectronic components is the shaft angle encoder as used in machine tool control and positioning equipment. Before the introduction of compact photocell arrays and infra red L.E.D's the encoder relied on wiping contacts sensing a conducting pattern on a special printed circuit board. The contact encoder had a relatively short life because of track wear. Its speed was restricted to a few RPM and it could normally be used in one direction of rotation only.

The use of optoelectronic components and a photo-graphically prepared tracked disc has extended the life and performance of the encoder to that of its remaining mechanical parts such as the bearings. They have also increased its speed of use to several thousand RPM and permitted rotation in either direction by the elimination of rubbing contacts. In some absolute encoders the accuracy and resolution has been increased by a factor of 4 by the use of very narrow slits of about 0.003" in front of the detectors to control the switching point on the coded track.

Another device whose performance has been improved by the use of optoelectronic components is the punched paper tape reader. Apart from the tape transport mechanism all other mechanical contact can be removed and the limit of operation is dictated by the strength of the tape and its ability to be driven at and stopped from high speeds. Reading rates have been increased from say 20 characters per second for mechanical reading to over 100 per second with optical reading, thus permitting more efficient use of computer terminal equipment, etc.

Both postal sorting and printed character recog-nition have been made possible by the use of light sensitive devices. It is difficult to envisage other than optical methods for these functions. Most customers of the Post Office would no doubt become very irate if holes were punched through their correspondence like paper tape to permit it to be sorted efficiently be mechanical means.

The self scanned optoelectronic detector array described previously can resolve a printed character on a cheque or account form into 120 elements and "read" each point in about 3 μS, or 200 μS for the whole character by using serial

and parallel scanning.   Therefore, if the information on each document to be read is printed in a
single row of say 50 characters, then the time taken
to read this row and transmit the coded information
to the associated computer is only 10mS.   It is in
such applications that great savings in time and
human effort can be made since a human operator
would probably need 10 seconds or more to complete
the same operation.   Also since the machine does
not get tired or bored with its task, as humans do,
its error rate is only 1 character in 100,000
compared with about 1 in 10,000 for a human operator.

Optoelectronic components offer longer operational
life up to 10 years or more and reduction of
servicing calls from 4 to 1 per year when used in
mechanical handling equipment.   Position sensors,
limit stops, go/no go tests, safety interlocks,
and movement sensors should all perform more
reliably over longer periods because of the
elimination of mechanical contact and fatigue
effects on springs and switches.   Initial costs of
the installation may be higher than mechanical
systems because of the associated electronics, but
the costs can easily be justified and offset against
reduction in service calls and machine down time.

Even the most delicate of mechanisms can be monitored
and controlled by optical means.   No force at all
is required to interrupt a light beam between an
emitter and sensor but the mechanism may not have
sufficient power to operate even the most sensitive
micro switch.   By the use of optical systems and
collimator slits with the detectors and emitters,
movements as small as 0.001" can be detected and
effect the switching of electronic circuitry to
give accurate control of the machine movement.

CONCLUSION

The use of optoelectronic components in mechanical
handling can improve equipment performance and
operating life but sometimes at increased cost.
These costs are invariably recouped by reduction of
servicing.   With certain problems they can offer
the only solution.   Their future is therefore
assured.

TABLE I

Comparison of L.E.D. characteristics

Material	Emission Colour	Peak nm	Optical Power output μW		Switching Speed	Efficiency typical
			DC	Pulsed		
GaP (N)	Green	550	8	40	30 nS	0.01%
GaP (N)	Yellow	575	25	200	30 nS	0.03%
GaP (N)	Yellow	575	200	2000	30 nS	0.05% High Power type
GaP (ZnO)	Red	690	200	1000	300 nS	1.0%
GaAsP	Red	660	90 μW	900 μW	10 nS	0.20%
GaAs	Infra Red	900	10 mW	300 mW	10 nS	0.8%
GaAlAs	Infra Red	900	20 mW	200 mW	10 nS	3.0%
GaAs (Si)	Infra Red	940	30 mW	500 mW	500 nS	5.0%

Fig. 1.   SIMPLIFIED CONSTRUCTION OF PHOTOTRANSISTOR

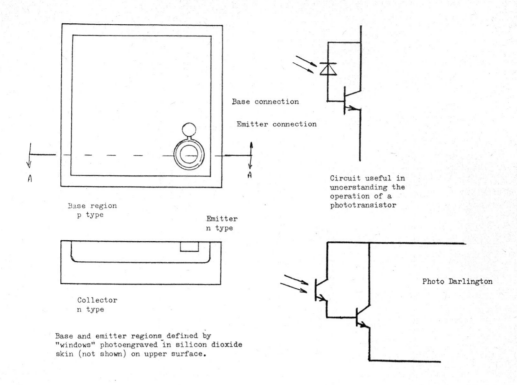

Base connection

Emitter connection

Base region
p type

Emitter
n type

Collector
n type

Base and emitter regions defined by
"windows" photoengraved in silicon dioxide
skin (not shown) on upper surface.

Circuit useful in
uncerstanding the
operation of a
phototransistor

Photo Darlington

Fig 2.   Normalised spectral response curves
of Solid State Detectors

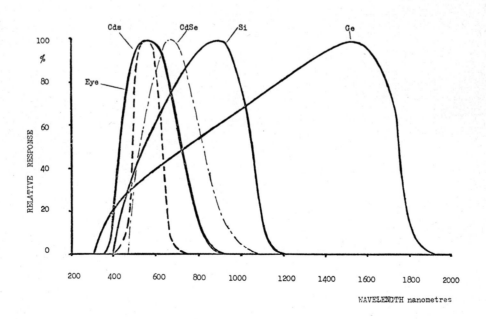

Fig. 3. Peak emission wavelengths of
L.E.D's with silicon photocell and
eye response

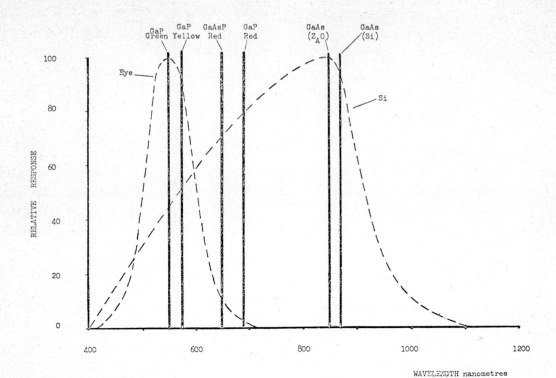

C146/74

LIGHTPEN SYSTEMS FOR IDENTIFICATION AND CONTROL

KENNETH DUNCAN FRASER CHISHOLM C Eng
ALAN J. SIVEWRIGHT B Sc
The Plessey Co. Limited, Sopers Lane, Poole, Dorset

SYNOPSIS   This paper describes the equipment modules used in Plessey data capture systems with particular reference to the Plessey lightpen and the bar-coded labels it reads.   This novel method of identifying articles is described together with a number of business applications which require articles to be identified and in some cases safeguarded.

BACKGROUND

The Plessey Company has developed a range of system modules intended for the collection of information about articles at source and its subsequent transmission to a data processing centre.

The information collected by the system is needed by the management of businesses for a number of purposes.   There is a need to locate and identify articles to control stock, for management is always anxious to keep stock at a minimum level.   Having identified and quantified stocks there is a need to reorder to ensure that stocks are maintained at the correct level.   This process of stocktaking and reordering sets many mechanical handling processes in action in manufacturing, warehousing and distribution.

The Company has also developed systems, the purpose of which is to safeguard articles which are relatively free to move.   By tagging the articles and tracing their movement through an organisation their whereabouts is continuously monitored.

METHOD OF TAGGING AND IDENTIFYING ARTICLES

The basis of these systems is a bar-coded label and a lightpen.   The lightpen is a lightweight cylindrical device superficially resembling a ball point pen (figure 1).   Essentially it consists of a light source and a solid-state photodetector element.   Light emitted by the source is collected by a cluster of optical fibres and transmitted along them to the tip of the "pen" where it illuminates a small area of label on which binary information is printed in the form of a "bar-code".

Light from this illuminated area is reflected back into the lightpen, where it generates an electronic signal in the photodetector element, corresponding to the bright or dark areas under the tip of the pen as it is scanned across the bar-coded label.

The bar-code consists of a series of wide and narrow black bars printed on a white background, the wide bars representing binary digit "one" while the narrow bars represent "zero".   Each label includes a 4-bit "start" code, and a

4-bit "reverse start" code which ensures that the label can be read either from left to right or from right to left.   Between these two codes are arranged the basic message which may be in 4-bit binary coded decimal and an 8-bit cyclic check code. This code structure is illustrated in figure 2.

In use, the lightpen is scanned along the bar code as if to draw a line, and provided the pen passes from end to end of the label without wandering off the sides, the data will be transferred correctly to the rest of the system logic.   The electronics have been specially designed to tolerate wide variations in the speed of scanning the code, to accept only words containing the correct number of bits and to reject any words without valid start codes or in which the cyclic check code indicates an error.   In the event of a label being incorrectly read, or of an error arising from any other cause, visible and audible warnings are generated.

The check method built into the logic circuits ensures that less than one error in a million will remain undetected by the equipment, and, therefore, if an operator misoperates the pen once on average in 100 reading operations, the undetected error rate would be less than one in 10^8.

The sensitivity and tip design of the lightpen have been arranged such that a satisfactory response can be obtained even when the axis of the pen is inclined at an appreciable angle from the perpendicular, so that the natural angle at which a pen is normally held, about 30^0 from the vertical, presents no problem.

LABELS

Labels can be reproduced by most printing processes. However, there is a problem in creating unique labels which are required for a number of applications.   To overcome the problem of setting up a unique label a label printer has been developed which produces bar-codes together with the associated human readable interpretation on adhesive backed labels.   This printer produces bar-codes of 32 bits per inch density and of the required quality to match the reading performance of the lightpen.

MODULES REQUIRED IN DATA CAPTURE SYSTEMS

Having captured data associated with a particular article this information must normally be passed on to a central processing centre for further action. Two basic systems used for collecting this data are an off-line system which records the data on a magnetic tape and an on-line system which uses a mini-processor.

OFF-LINE EQUIPMENT

Figure 3 is a block diagram showing the elements used in an off-line system.

PORTABLE DATA CAPTURE UNIT

The first unit is a portable or mobile data capture unit which has a lightpen for reading fixed data from bar coded labels. Part of this unit is a keyboard which can be used to enter variable data like quantities. It can also be used to enter the fixed data if, for example, a label is mutilated. If fixed data is being keyed in, the unit has a check-digit feature which ensures that only correct numbers are entered. The display shown is used to indicate the numbers which have been keyed in or entered by the lightpen. The display also serves as an alarm indicator and it flashes, for example, if a label is incorrectly read. All the data captured by the lightpen or through the keyboard is recorded on a small cassette - similar to a cassette used in a domestic cassette recorder. Being portable the unit is powered by batteries which are recharged when the unit is connected to the data transmitter or a separate battery charger.

DATA TRANSMITTER

Once a recording session is completed the data on the cassette must normally be sent to the computer centre. This may be done by removing the cassette from the portable data capture unit and manually transferring it to the computer centre. If, for example, this process takes too long then the cassette is not removed from the data capture unit, but the whole unit is taken to a data transmitter to which it is connected by means of a plug at the base of the portable data capture unit. The data transmitter is connected to the computer centre through the Public Switched Network via a modem. The computer centre telephone number is dialled and once a connection has been made the transmitter is switched to transmit and the data held on the cassette is sent over the public telephone lines to the computer centre. Alternatively, the transmitter can be left unattended and when the computer centre dials the telephone number of the telephone to which the transmitter is connected, an auto-answering unit will accept the call and the data on the cassette will be automatically transmitted.

RECEIVER UNIT

A cyclic check code similar to that used in lightpen reading is generated and added on the end of a data block recorded on the cassette. The receiver checks the data with the cyclic code and if an error is detected a message is sent back to the transmitter requesting retransmission. If the telephone call is interrupted the cassette is automatically rewound to the beginning to await another call from the receiver. Up to sixteen telephone lines can be handled simultaneously by the receiver. All the incoming data is recorded onto magnetic tape in a suitable format for the central computer.

DATA CONVERTER

This unit (not shown on figure 3) takes cassettes which have been sent in, and converts the data on these cassettes onto computer compatible magnetic tape in much the same way as the receiver.

ON-LINE EQUIPMENT

Figure 4 illustrates a typical on-line system.

COUNTER TERMINALS

These units are very similar to the portable data capture units in that they have a lightpen, keyboard and display; they have additionally, however, another group of indicators which can be operated by the central computer to give two-way communication. An alphanumerical printer can be connected to the terminal if hard-copy is required. In addition, a visual display unit with full alphanumerical keyboard can be incorporated to provide extended operator facilities.

TERMINAL

A number of remote data capture units can be handled in one location by connecting them to a multiplexer. Included in this unit is a remote transmitter module which sends the data to the computer centre and handles the data streams in much the same way as the off-line system. The transmitter can send data over lines leased from the Post Office or over the Public Switched Network.

CENTRAL COMPUTER

A mini-processor is used to collect all the incoming data, to control transmission and to check for invalid data blocks and incorrect message formats. It will signal back to the appropriate terminal if there is an error and it will also send back other information destined for a terminal. All the incoming data can be buffered on computer compatible magnetic tape for later processing by an E.D.P. computer, or the mini can act as a front-end machine to a host central processing unit.

APPLICATION OF DATA CAPTURE SYSTEMS

With these tools a large number of problems can be solved in a more effective way than with earlier solutions. A number of applications will be described to show the versatility of this equipment in the general categories of Identification and Safeguarding. .

IDENTIFICATION

Product ordering

A national chain of supermarkets uses three large warehouses as depots to feed individual shops. Deliveries are made every night and their problem is to let the depots know what goods the shops need so that they are fully stocked to carry out their next days trading. At the beginning of each day the shelves are stocked but as shoppers take goods out, the shelves become depleted. Certain bulky lines like flour and sugar are held in a back room and are brought forward as shelf space becomes free. The majority of the lines however are held on the shelves. Once a shop is closed an operator can walk round the shelves and decide which lines need reordering. A bar code label is fixed to the shelf edge identifying the product, so by wiping the lightpen on the portable data capture unit across the label a case quantity is reordered. Two wipes orders two cases, and for greater orders the quantity is entered using the keyboard. At the end of an ordering session, which takes about an hour, the portable data capture unit is attached to a transmitter and left unattended. A receiver at the computer centre polls all the shops and transfers the order data onto magnetic tape. A central computer sorts all the orders into picking list order per shop and transmits these picking lists to a line printer in the warehouse. Here, fork-lift truck operators collect the cases and load them into lorries. When a lorry is full it is driven to the shop through the night and early morning and the goods are unloaded and stacked on the shelves ready for trading. The Plessey data capture system is the key which opens the door to this enormous warehousing and distribution service.

Stock taking

Another use of the same set of equipment is to take a physical stock count of items in warehouses and shops. This information is needed to reconcile actual stock with what the central computer thinks is held at each location.

Warehousing and distribution

One of the problems of warehousing is knowing what is held in stock and where the stock is located. When articles are manufactured by a sub-contractor they are supplied with a set of pre-printed bar-code labels. Alternatively, when articles arrive at goods inwards they are given an identifying label. From a stock file the warehouseman is given an address of a bin location. The goods are taken to the bin and the bar-code of the bin address is read together with the bar-code of the goods, so that the two are correctly linked together. This information is fed to a computer which consolidates the location files. Orders on the warehouse are readily translated to bin locations and deliveries are assembled at goods outwards. Here the bar-code label of the vehicle taking the goods is read together with the identities of the articles being loaded into the van. Once a van is loaded it is locked and the central computer is given a complete record of everything being sent to a particular destination.

In some applications it is not possible to label individual items in the warehouse. In this case a message can be sent to the receiving shop where there is a bar-code label printer and labels can be produced in the shop in anticipation of the goods just sent from a warehouse.

Stock control

A new concept in retailing recently introduced, figure 5, uses a catalogue to describe the goods being offered. The goods are held in warehouses throughout the country in which there are display cases with a selection of goods on display. The goods themselves are racked in bins in large storage areas which can extend to several stories in height. Customers select the items they want to purchase from their catalogue and approach the sales counter where there are a number of lightpen terminals. Each sales girl has a special catalogue with bar-coded labels beside each item. An article is ordered by wiping the appropriate bar-code label and if it is in stock a printer prints a label showing the product number. The printer is located at the nearest point to the stock. An operator picks up the ticket, searches the racks, collects the article, sticks on the label and puts the parcel on a conveyor to the goods-out position. Meanwhile, the customer has paid for the goods and by the time this transaction is completed his goods are waiting for him. If an article is out of stock a signal is sent back to the terminal allowing the customer the option of changing to another product.

Another variation on this theme is with a plant hire firm. In this case the data captured is the customers name and address which is entered through a keyboard on a visual display unit, and the product they want to hire which is entered by a lightpen reading bar-code from a catalogue. This data is sent via the Public Switched Network to a central computer which checks whether the item is already on loan. If it is, a list is sent back to the visual display indicating who has the item and the due-back date. If it is not out on loan, a contract is sent back and printed out on a small line printer.

SAFEGUARDING

In a number of applications it is important to safeguard articles or documents as these move from place to place. Basically, this is achieved by automatically recording the identity and location of material as it moves around an organisation.

Document tracing

Confidential files are registered by entering their title and bar-code reference number into a computer. Throughout an organisation there are a number of simple lightpen and keyboard terminals connected on-line to a central mini computer. A document is issued and is received by wiping an identity label (identifying the person receiving the document) and the label on the document. To pass the document on to another person the identity label and document label is read again and the address of the destination of the document is keyed in. The central computer can thus track the progress of documents, raising alarms if a document is not received in time and it can also build up dossiers

on the material passing through the hands of an
individual.

Parcel tracing

A similar approach is used to safeguard parcels
carried by a security firm.   In this case, when
a parcel is collected it is given a unique label.
When the parcel is loaded into a van its bar-code
label is read together with the van route number.
Whenever parcels are removed from vans and resorted their numbers are read together with the
route number of the new van.   All this data is
sent to a computer centre which tracks the progress of parcels in and out of the network.
Parcel numbers which have not cleared the system
before a specified time are indicated to a super-
visor so that internal chasing action is initiated
before the customer notices!   An enquiry system
is available to check the progress of any particular parcel.

LIBRARIES

One of the most useful applications of the lightpen
is to safeguard books in public libraries
(figure 6).   Here, books, records and paintings
are all given a unique label.   To issue material
the borrower's number is read followed by the
numbers on the books being borrowed.   A date
stamp attached to the lightpen is used to stamp
the date when the books are due back.   To return
books only the book numbers and not the borrower's
number are read. .   All this data is sent to a
computer centre which keeps a record of all the
books out on loan and their associated borrowers.
If a book is kept beyond the due back date the
central computer can automatically send out a
reminder notice.

In addition to the circulation control described
above, by adding a small core store to the branch
equipment, books can be trapped on their return
which are reserved and dilinquent borrowers can
also be trapped.

The modules which have been developed can cater
for a very small branch library, a mobile
library, a central library and a group of large
libraries in a town or borough.   A number of
universities are also using this system.

CONCLUSION

Bar-code reading techniques are becoming accepted
and will be supplied in increasing numbers to a
variety of industries to assist in business management. As businesses become larger and complex these
systems will help to identify and safeguard goods
and to control the interaction between buying and
selling, storage and transportation, and provide
valuable management information.

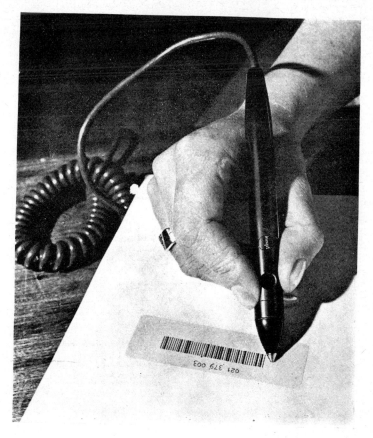

Fig. 1.   The Light Pen

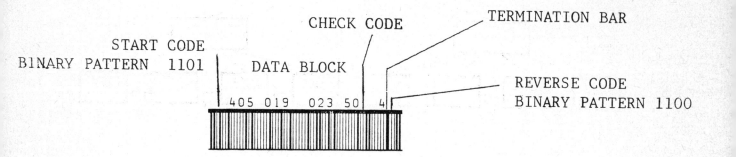

Fig. 2.   Typical Label Detail

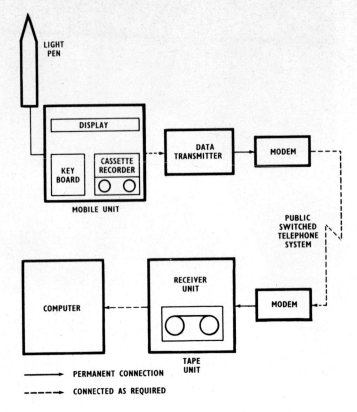

Fig. 3. Off-line System

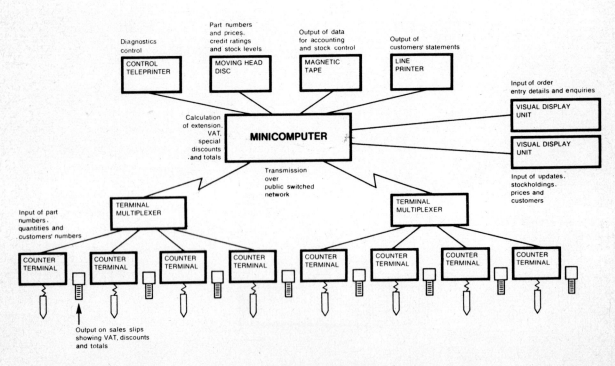

Fig. 4. Typical distributive trades point-of-sale system

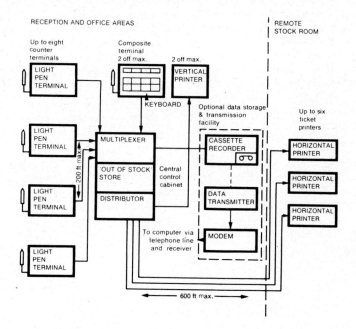

Fig. 5. Typical closed trading system

Fig. 6. Photo of Library System

List of Delegates

Adey, S. J.	Ever Ready (Great Britain) Ltd, London
Agate, M. S.	Postal Headquarters, London
Andrews, J. D.	Post Office, London
Armstrong, J.	Roberts and Armstrong (Engineers) Ltd, Wembley, Middlesex
Axon, J.	Post Office, London
Bailey. N. G.	Postal Headquarters, London
Barr, D. J.	Post Office Research Centre, Martlesham
Bartlam, P.	Integrated Photomatrix Ltd, Dorchester
Bates, R. V.	Postal Headquarters, London
Beal, D. W.	The Diamond Trading Co. Ltd, London
Bennett, I. N.	Ilford Ltd, Brentwood, Essex
Bennison, R.	Cambridge Consultants Ltd
Bielby, R. W.	Imperial Metal Industries Ltd, Birmingham
Bordes, M. P.	French Post Office, Paris, France
Bossons, W. H.	Masson Scott Thrissell Engineering Ltd, Bristol
Brewster, I.	I.C.I. Plastics Ltd, Welwyn Garden City
Bright, D. E.	Post Office, London
Brimelow, P.	Bank of England Printing Works, Loughton, Essex
Broadbent, S.	Harrison and Sons Ltd, High Wycombe, Bucks.
Broido, D.	Mackintosh Consultants Co, London
Buckle, G.	Bank of England Printing Works, Loughton, Essex
Busby, J. L.	Postal Headquarters, London
Cadwallader, M. V.	Burroughs Machines Ltd, Croydon, Surrey
Carpenter, H. J. T.	G.E.C. - Elliott Mechanical Handling Ltd, Erith, Kent
Chandler, K.	Masson Scott Thrissell Engineering Ltd, Bristol
Chapman, E. J.	Post Office Research Centre, Martlesham
Clark, W. J. R.	Marconi Elliott Avionic Systems Ltd, Basildon, Essex
Clarke, B. S.	Cranfield Institute of Technology, Cranfield, Beds.
Copping, G. P.	Postal Headquarters, London
Curl, B. J.	De La Rue Instruments Ltd, Portsmouth
Davey, D. V.	Postal Headquarters, London
Davies, A. O.	Plessey Co Ltd, Poole, Dorset
Elsworth, J. F.	Kenrick and Jefferson, West Bromwich
Fairhead, M. J. B.	Letraset International, Ashford, Kent
Fellows, E. C.	Crabtree - Vickers Ltd, Leeds
Fletcher, J. H.	Post Office Research Centre, Ipswich
Galea, S.	Postal Headquarters, London
Gibbs, I.	Sovex Ltd, Erith, Kent
Giles, A. F.	Unilever Ltd, London
Gloeckner, J. E. P.	Lippke (U.K.) Ltd, Windsor, Berks.
Goddard, F. A.	Postal Headquarters, London
Godfrey, S. W.	Post Office, Birmingham
Goodison, H.	Post Office, London
Greathead, T. W.	Postal Headquarters, London
Haigh, L.	Needle Industries Ltd, Poole, Dorset
Hall, R. C.	Post Office Research Centre, Ipswich
Hall, R. J.	Wiggins Teape Research and Development Ltd, Beaconsfield, Bucks.
Harrison, J. C.	Post Office, London
Hearn, E.	Postal Headquarters, London
Henly, H. R.	Postal Headquarters, London
Hewett, J. W.	Postal Headquarters, London
Hills, E. G.	Postal Headquarters, London
Hodgson, S. C.	Kenrick and Jefferson Ltd, West Bromwich
Holloway, P. I.	G.K.N. Screws and Fasteners Ltd, Smethwick, Staffs.
Holt, J. B.	National Institute of Agricultural Engineering, Silsoe, Beds.
Hornsby, H. C.	Post Office, Brighton, Sussex
Howell, J. S.	Plessey Co Ltd, Poole, Dorset
Hutchinson, P.	Glaxo Laboratories Ltd, Greenford
Jacob, N. L.	Walmore Electronics Ltd, London
Janzen, N. P.	Timsons Ltd, Kettering, Northants
Jenkins, S. M.	De La Rue Instruments Ltd, Portsmouth
Keehan, M. J.	Frazer Nash (Electronics) Ltd, Kingston upon Thames, Surrey
Kempson, J. H.	G.E.C. - Elliott Automation Ltd, Leicester
Langley, G. B.	Post Office Research Centre, Martlesham
Lee, W. G. H.	Post Office Research Centre, Martlesham
Lutteman, S. M.	Harrison and Sons Ltd, High Wycombe, Bucks.
Lyall, A. R.	Letraset International, Ashford, Kent
McCarthy, C. R. S.	Postal Headquarters, London
McGill, R. J.	Wiggins Teape Ltd, Aberdeen
Matthews, P. C.	Plessey Telecommunications Research Ltd, Poole, Dorset
Maughan, W. S.	Gunsons Sortex Ltd, London
Mellor, D. C.	National Materials Handling Centre, Cranfield Institute of Technology
Milne, F. A.	Post Office Research Centre, Martlesham
Nash, P.	N.A.I.S. Ltd, London
Newman, T. E.	G.E.C. - Elliott Mechanical Handling Ltd, Erith, Kent

Nicholson, F. J.	Ministry of Agriculture, Fisheries and Food, Aberdeen
Nokes, L. A.	Post Office, Birmingham
Norman, B. W.	Kodak Ltd, Harrow, Middlesex
Oatey, L. W.	Post Office, London
O'Doherty, P. J.	Post Office, Brighton, Sussex
Osborn, H.	Wiggins Teape Research and Development Ltd, Beaconsfield, Bucks.
Owens, C. R.	Mars Ltd, Slough, Bucks.
Parker, B. W.	De La Rue Instruments Ltd, Portsmouth
Peacock, A. R.	Post Office, London
Pettit, A. G. R.	Post Office, London
Phillips, K. H.	Post Office, London
Pilling, T.	Post Office, London
Polydorou, T.	Ever Ready (Great Britain) Ltd, London
Potter, N.	Post Office, London
Price, P. R.	Consultant to the Printing Industry, Epping, Essex
Prior, V. M.	Marconi Elliott Avionic Systems Ltd, Basildon, Essex
Purll, D. J.	SIRA Institute, Chislehurst, Kent
Read, P. G.	The Diamond Trading Co Ltd, London
Richardson, D. J.	Data Dynamics Ltd, Hayes, Middlesex
Robin, B.	Mars Ltd, Slough, Bucks.
Rooks, B. W.	University of Birmingham
Sanders, H. H. A.	Thorn Automation Ltd, Rugeley, Staffs.
Selwood, J.	Ministry of Defence, London
Sharpe, B. A. J.	De La Rue Instruments Ltd, Portsmouth
Simmons, D. A.	Post Office, London
Smith, E. B.	Fife Controls, Redditch, Worcs.
Smith, G. L.	Post Office Research Centre, Ipswich
Smith, J.	Post Office, Manchester
Smith, J. Q.	Crane Packing Ltd, Slough, Bucks.
Smith, M. G.	G.E.C. - Elliott Mechanical Handling Ltd, Erith, Kent
Smith, M. H.	Roberts and Armstrong (Engineers) Ltd, Wembley, Middlesex
Spiller, B. H. R.	Decca Radar Ltd, Walton-on-Thames
Styan, P. O.	Plessey Co Ltd, Poole, Dorset
Tarbet, A. G.	Post Office, Edinburgh
Tennant, D. J.	Post Office Research Centre, Ipswich
Thomas, A. F. T.	G.K.N. Screws and Fasteners Ltd, Smethwick, Staffs.
Thomson, M. G.	Manufacturers Equipment Co Ltd, Hull
Thomson, P. G.	Ever Ready (Great Britain) Ltd, London
Tottey, S. T.	G.E.C. - Elliott Mechanical Handling Ltd, Erith, Kent
Turton, P.	Plessey Co Ltd, Towcester
Wakefield, J.	Postal Headquarters, London
Walker, A. D.	Post Office, London
Watts, P. E.	Vickers Ltd, Ascot, Berks.
Weeks, T. N.	Post Office, London
Whitchurch, H.	Post Office, London
White, I. G.	Post Office, London
Whiter, D. W.	Postal Headquarters, London
Whittington, K. W. H.	Postal Headquarters, London
Wicken, C. S.	Post Office, London
Wilson, J. W.	Post Office, London
Wood, D.	Masson Scott Thrissell Engineering Ltd, Bristol
Wood, T. D. S.	Post Office Research Centre, Ipswich
Woodward, C. A. W.	National Engineering Laboratory, East Kilbride
Yeadon, N. B.	Post Office, London
Young, G.	Needle Industries Ltd, Studley, Warwickshire

Subject Index

Titles of papers are in capital letters